175 More Wow! Change Cards

Leaders Inspiring Others During Change

Christina M. Battell

and

Deborah C. Miller

DISCLAIMER / LEGAL NOTICE

ISBN: 978-1-939192-13-4 [Paperback]

ISBN: 978-1-939192-14-1 [Kindle]

Table of Contents

examples can show where the organization can improve change efforts.

Variation 8...Ask prospective training participants to choose one WOW! Change Card that you would like him/her to present during training. The presentation can be simple. Read the card. Comment on the card. Offer a real-world example of the card's lesson. The real-world lesson could be personal as well as professional.

Ask participants to pick a WOW! Change Card to share with his/her team, department or colleagues. Ask each participant to bring comments from those people to class, along with comments about change management in the organization.

II. Rev-Up Classroom Training

Use WOW! Change Cards to start training with a bang. Make introductions interesting and worthwhile. Have each participant pick a card randomly from the book (this would be particularly effective using cards with quotes). Ask the participants to consider the card's meaning and share that in the introduction.

Use the WOW! Change Cards for small group discussion. Ask each small group (2-8 participants) to select a card randomly from the book. The small group should read and discuss the card. They should then report key points to the whole class.

Variation 9...Each small group should be assigned the same card so discussion results can be compared across the whole class.

Variation 10...Each small group will receive a card or several cards to read. Then, ask the small groups to discuss how effectively the organization fulfills the requirements of the card, with examples as evidence.

Variation 11...The assessment might include a rating scale of 1-10; 1 = poor, 10 = excellent. Each group should offer a rating for each assigned cards, offering comments and examples to support their rating.

Create discussion tables. Assign a facilitator for each table. Ask small groups to rotate from table to table. The facilitator should remain at the same table for all discussions. Each small group should add to the discussion notes during each round. Facilitators should share and post results at the end of the small group discussions.

Energize the group. After participants return from a break, read a card and ask for several brief comments before digging back into the agenda.

Conduct a group grope (Yes, we said "grope"). Pass out copies of 3 cards to each participant. Ask the whole group to mix for 10 minutes, comparing cards and trading cards when they wish. Ask them to talk with as many people in class as possible during the 10 minutes. Put some cards on a side table for additional trades (go to http://www.thiagi.com for more information on Group Grope).

Give a book to each small group of 5-8 participants. Ask each small group to select 5-10 cards that they find most interesting. Ask them to discuss why those cards draw their attention. Then, ask them to prioritize the cards according to level of interest or importance. Each group should report their results to the class.

Use WOW! Change Cards for classroom transitions. Before breaks or at the end of a training day, ask small groups to select a card that conveys a key learning point. Ask each group to share the card and why they chose it.

Break up the classroom action by asking a participant to choose a WOW! Change Card from the book. Take a few minutes to read and discuss the card before moving to the next agenda item.

III. Make Training Stick!

Reinforce training concepts, ideas, and actions by sending a WOW! Change Card book with each participant. Explain how to use the cards for on-going learning after classroom training.

Ask training participants to submit a post-training report about how each used a change concept or idea in life or work.

Pick a WOW! Change Card that each participant should read and review after training. Ask that the participants submit an email or internet/intranet report explaining how he/she used the card concept or idea.

Send each training participant a copy of a WOW! Change Card each week for at least six weeks. Ask for feedback about how each participant has used the card concept or idea.

Variation 1...Select discussion cards before the meeting. Choose cards most relevant to your particular change initiative.

Variation 2...Assign the same card to each small group. Ask each small group to report to the large group. Compare the key points from each small group.

Do a reality check. Ask participants to discover real-world examples of the change card concept or idea in action—successful or otherwise. Make the assignment at this meeting, giving time to gather real-world examples before the next meeting.

Play a change game. Pass out copies of change card pages to meeting participants. The cards can go to individuals, pairs or small groups. Ask each card holder to study the card. Then, each must apply the card lessons BETWEEN the current meeting and the next. At the next meeting, all can compare notes about experiences applying lessons.

Do a change assessment. Pick up to 10 WOW! Change Cards. Divide the meeting participants into pairs or small groups. Give the "10 pack" to each set of pairs or small group. Ask each pair or small group to assess the effectiveness of a current change initiative in each of the 10 pack areas. Each pair or group should report discussion results.

Variation 3...No time to assess and report this time? Make the assessment assignment today and ask for reports at the next meeting.

Variation 4...Instead of a "10 pack," give each pair or small group fewer cards. Even one card will do.

Variation 5...Instead of pairs or small groups, ask individuals to do the assessment and report results.

Variation 6...Set an assessment scale such as 1-10; 1 = poor 10 = excellent. Ask for reports that offer both an assessment item rating and explanation.

Ideas For Training

I. Get Ready For Training

Use WOW! Change Cards for classroom training preparation. Give each participant a book for the pre-training work. Ask participants to select 3 cards before coming to training. One card should describe a success or strength of the organization during change. One card should represent a change challenge of the organization. The third card should represent one question the prospective trainee would like to have answered. Ask for cards to be shared during training introductions or during strategically placed small group sessions.

Shorter training? Ask for just one card to be selected during pre-work. Hear about the cards during introductions or during a small group activity.

Variation 7...Ask participants to search for real-world examples illustrating the card of their choice. The examples can illustrate successful change efforts by the organization. Or, the

PICK A CARD...ANY CARD

This book began as several sets of cards, as we found decks of cards convenient for classroom training. We found plenty of creative ways to use cards with both individuals and groups. Cards are versatile, portable, easy to handle and quick to digest, making them ideal for classroom and desktop use.

What we also discovered is that book format makes for easier printing. So, what started as decks of cards has evolved and now you have a book in front of you instead of full decks of cards.

Please keep this in mind as we use the term "card", even to this day, and despite the fact that we are printing books, not actual cards. Yes, you will literally turn "pages" in this book, but please note that "pages" are actually "cards" of the past. Hold that thought while you read the recommendations for how to use "cards". Here they are:

Ideas For You

Keep the book close at hand. Focus on one card (lesson) per day. Read the card several times during the day. Reflect on what the change lesson means to you. Start a change journal and write your answer to the discussion question at the end of each card.

Not a very methodical person? No matter. Pick your "daily dose" of change from anywhere in the WOW! Change Card book. The fun is in the surprise. Read the card several times during the day. Reflect on what the change lesson means to you.

Start with the big picture. Scan the whole WOW! book. Make note of the most interesting or relevant sections. Use those cards first. Do the others later, based on interest or need.

Scan the WOW! book, looking for 3-5 lessons to reinforce recent change training that you attended. Copy and carry those cards as reminders of what or how to apply change lessons on the job.

Ideas For Meetings and Teams

Use the WOW! Change Cards at meetings. At regular staff, department or project meetings, ask a participant to pick a page in the book—randomly select or draw a number from a hat will work. Ask another participant to read the card. Lead a discussion of the question on the card.

Want to involve more people? At the end of this meeting, ask for a volunteer to select a change card for the next meeting. Ask that person to also be in charge of leading the discussion. There will be plenty of time for him or her to prepare between this meeting and next.

Have a larger meeting? Break the large group into smaller groups (5-6 participants per group is nice). Ask each group to pick a card, read it, and discuss it. Have plenty of time? Ask small groups to report back key points to the large group.

mixing weird concoctions and cures, even tried out a few of their newly discovered potions for prying loose of the status quo.

They experimented, admittedly with some failures. After each set-back, the two picked themselves up, continued the research, reading, and discussion and continued on their way. Sometimes, there were side trips to locations just off the beaten path of change, but still quite related to it—mini-adventures inside the bigger change adventure. And, there were many successes which seemed very much like moments of a trip when one arrives at an astonishing site or discovery, steps back in awe, delight, and maybe even takes a photograph or two to make a memory for future reflection.

The result of all the activity? Two books. The first book, ***125 Wow! Change Cards: Tools For Leaders During Change*** leads the reader through the challenges leaders face and the human response to change. In this book, ***175 More Wow! Change Cards: Leaders Inspiring Others During Change***, we explore the many ways leaders can light a fire for change.

Introduction

Chris Battell and Deb Miller are friends and colleagues who share a love for adventure. Learning provides one of their favorite types of adventure. So, Chris—who works with leaders, teams, conflict, meetings and the like—and Deb—whose specialties are projects, efficiency, technical writing, and process improvement (amongst other things)—took on the adventure of studying change—both the literature and real life. The two noticed that their own success depends on the ability to change a variety of things, including themselves. They also saw that their clients needed the same ability to change—or run the risk of being left behind the rest of the more nimble souls who populate workplaces and the world.

With a destination clearly in mind—discover the best and proven ways to change and change for good—the two embarked on the journey. They traveled through the contents of many books about change. The two of them visited with clients about change. They made observations about change and its impact in their personal and professional lives. They talked with colleagues, read blogs and participated in conversations about change via social media.

Chris and Deb also ran some experiments. They used change principles and practices with clients. And, like Mad Scientists

1 Inspire Change: Perceptions Vs. Reality

When A Snake Is Really A Stick

There once was man who had a run-in with a snake. He escaped without harm, but never forgot what it was like to face the fearsome creature. His heart raced. His pulse surged. His body stiffened as he readied to act! The man was wired to survive and what a good thing indeed.

Some years later, the same man walked down a wooded path with a loved one. He was relaxed. In bliss. Until he saw IT! A stick in the path. To the man the stick looked like the snake from long ago. His heart raced. His pulse surged. And then he saw what was really there.

A Stick. Not A Snake!

It can happen to the best of us. We see something and mistake it for something else. The co-worker who appears to be gossiping—about us—but is really pouring her heart out to a friend. Yes, the tones are hushed…but gossip? No. Or, there's the boss who appears aloof, indifferent to all those around him or her, but is really preoccupied with today's emergency. Is the inattention caused by any of the people present? No, not in the least.

Be careful that your perceptions don't get the best of you. Certainly perceptions matter, but so does reality. So, when emotions run high, do a perceptions vs. the reality check. The questions are:

How might my past experiences color the changes we face?

How does my perception compare to the facts?

How do the perceptions of others compare with the facts?

2 Inspire Change: The Psychology of Moving People

The Power of Pleasure and Pain

How to get people moving? Employ the tool that Kouzes and Posner (The Leadership Challenge) call "the vision thing." In Kouzes and Posner terms, describe "a vision of an uplifting and ennobling future." In other words, talk about the positive, long-term picture of what can be accomplished through the change. While you're at it, compare the future ideal with the current reality, pointing out the costs of maintaining the status quo. The sharp contrast between the scenarios—the pleasures of an uplifting future and the pain of the way things are done now—should convince people that the current way isn't the best way.

Let's take the very challenging change—quitting smoking.

What is the Pleasure?

- Deeper, freer breathing
- Increased energy and stamina
- Fatter wallet
- Sweeter smells
- Healthier family and co-workers
- Thriving love life

What is the Pain?

- Coughing, wheezing, heavy breathing
- Fatigue and low energy
- Money out the window
- Hold your nose please
- More illness all the way around
- Who needs love? I've got smokes!

What are the pleasures of a current organizational change?

What are the pains of the status quo?

3 Inspire Change: Critical Thinking

A Test For The Critical Among Us...

Critical thinking is critical. But when and how?

The thing is, some of the best critics—those endowed with the indispensable gift of finding the flaws in a thing or the potential pitfalls of an idea—sometimes also take criticism too far. Too much. Too often. Yes, a critical view helps to perfect a thing or to avoid needlessly stumbling on the journey ahead. But to keep criticism from becoming a discouraging force, how about pointing out the good stuff too…and FIRST?

If 95% of something is fine, point out the 95% first. That, the sting of hearing about the 5% is not so tough.

What 95% is right with the thing?

What 5% could use work?

YOUR NOTES:

4 Inspire Change: Critical Thinking

Critical Thinking Or Just Critical?

The world and your workplace could use more of one and less of the other.

Critical Thinking

- The ability to weigh pros and cons
- A desire to get a thing done right by examining it from all angles
- A spirit for excellence and improvement
- Attacking problems not attacking people
- Careful judgment based on equally careful analysis

Just Critical

- Focus on flaws
- A compulsion to correct, fueled by a desire for superiority or control
- Criticism motivate by a desire for one-upmanship and domination
- Attacking people, not problems

- A rush to judgment, lacking careful consideration of the entire situation or the big picture

Which version of being critical do you bring to change?

5 Inspire Change: Two Paths To Change

Rational Or Emotional?

When faced with change—and the need to convince others about the value of the change—you can make the rational case, the emotional one, or both. We suggest you do both.

The rational case appeals to the analytical mind with logical arguments for why change is necessary and what it can accomplish. The rational case offers facts and data, showing that a new way of doing things will outperform the old way in some fashion that makes the challenges of change worth it. Change experts often call this the Business Case for change.

While the business case for change is essential—who wants change for change's sake after all—the emotional case for change powers up people's motivation for going forward. The emotional case for change paints a picture of how the change will impact lives and work at a feeling level. For example, the emotional case for change outlines how the change will make things easier, exciting, pleasurable, competitive, rewarding, attractive, successful and so on—all appeals of a more personal nature. The emotional case for change links the change with people's hopes, dreams, and wishes.

The difference between the logical and emotional cases for change is the same as the difference between a feature and a benefit. The icemaker on a refrigerator may produce ice—a nice feature—but isn't the person fixing a drink more interested in how easily and quickly one can dispense ice, with just one hand, no less? Features are the mechanics. Benefits are the emotional pay-off. Both are important to change.

What is the rational case for the current change?

What is the emotional case for the current change?

6 Inspire Change: Make the Case

Making The Case: Not For Lawyers Only

You know the importance of making the case for change, but how to do it?

Gather facts and data to make the rational case for change.

- What are the performance gaps between today's reality and tomorrow's potential?
- What are the current metrics? Revenues? Expenses? Other financials? Market Share? Competitive position? Workforce stats?
- What are the projected metrics, once change is accomplished?
- What are the organization's strengths and weaknesses today?
- What are the opportunities and threats—external challenges—making change urgent?
- How will the proposed change impact each of the above?

How many of these questions can you answer?

What are those answers?

YOUR NOTES:

7 Inspire Change: Make the Case

The Softer Side Of Making The Case For Change

The emotional case for change may be considered soft—and therefore weak—by some, but don't be fooled.

The soft side of the case for change—skillfully delivered—packs a big persuasive punch. The emotional case for change inspires people, convincing them to overcome fears of change, exciting them instead about the potential of it. The emotional case for change moves people beyond merely understanding the facts that prove the value of change to releasing the energy, hope, enthusiasm, and inspiration that will make the change happen. Construct an emotional case for change by answering these questions:

- What is our reputation—in the community, with key stakeholders, and with the competition?
- How do our customers experience our products or services?
- How are employees affected by how we do business today?
- How could we work differently once change is accomplished?
- What will the workplace look like if we change? How will things be easier, more satisfying, rewarding, successful, more efficient, more cost effective, etc.?

- What were our biggest failures last year and what difference did those failures make in people's lives? The community?
- How will those failures be transformed by the change?
- When we poll customers for testimonials and feedback, what do we hear?
- What will happen if nothing changes?
- What if the change is successful?

How many of these questions can you answer?

What are those answers?

8 Inspire Change: Innovation

Innovation Or Kaizen?

Breakthrough change and change accomplished one small step at a time each have a place in this world...and in business.

Big, sweeping, more dramatic change can be remarkable. That's easy to see. But the change method called Kaizen—incremental change, accomplished one small step at a time—can be remarkable too, and with certain advantages. While on the one hand, innovation typically demands shocking, more radical reform, Kaizen is founded on small, more comfortable steps towards improvement. The result? Less resistance but potentially incredible results. The lesson of Kaizen is simple. Small steps added up over time create big results and with less rattled nerves on the part of the people.

Don't have an hour a day to exercise? Spend 2 minutes on the treadmill. Walk the stairs instead of taking the elevator. Small steps are better than no steps.

Want to manage stress? Take one deep breath. 15 minutes of meditation are nice but are not necessary...today. Maybe tomorrow one deep breath can become 10 deep breaths.

Need to clean up your desk? Set a timer for 5 minutes and clean for that amount of time only. That's it. Enough for today. Once you're moving, 5 minutes may become 15.

The beauty of Kaizen? It's change that goes AROUND our brain's natural tendency to fear anything that seems a bit too different. The change related to Kaizen is too small to arouse fight or flight, and in that fact lies a distinct advantage. Small steps really can be the way to big progress. Use them.

What small step can you make today toward the achievement of a desired change?

9 Inspire Change: Start Strong...

Every Day

Super stars start strong. They treat the first 9 seconds of any event—the day, the project, the meeting, the game—with the utmost care. You should start strong too.

One smart way to do this: start the day off right. The moment you arise—and for the first 9 seconds right after that—set a vision for the day. What do you believe is possible? What do you want to achieve? How do you want to feel? See the answers to the questions in your mind's eye. And, see the possibilities of the day as if every bit of it is happening already. Savor the picture. Then, take that lovely picture of the possible with you—along with the enthusiasm it evokes—for the rest of the day.

Or try this exercise, suggested by Kouzes and Posner in *The Leadership Challenge*. Treat each day as if it was the first day on the job. Bring the same level of excitement, enthusiasm, optimism and dreams as you had the first day of the job to your job today. How's that for pumping yourself up?

Or, here's a motivating mental work-out for the day's start from *Change the Way You See Everything*. Visualize the day to come in the most positive, motivating, empowering and strengthening terms. Again, a vivid picture helps. Then, repeat this mantra 3 times (the 3 times matters):

I am the perfect person to accomplish (your vision), because I have (your talent or strength).

10 Inspire Change: One Step At A Time

Chunk It Down And Move It Along...

One step at a time is all it takes. No giant leaps necessary. Start now.

When you improve a little each day, eventually big things occur. When you improve conditioning a little each day, eventually you have a big improvement in conditioning. Not tomorrow, not the next day, but eventually a big gain is made. Don't look for the big, quick improvement. See the small improvement one day at a time. That's the only way it happens—and when it happens, it lasts.

~John Wooden, basketball coach

The Japanese call it Kaizen, otherwise known as continuous improvement. While the Japanese made Kaizen famous (ever heard of Toyota?), the practice was originally taught by the U.S. government to Corporate America in *Training Within Industries* classes. The time? World War II. The mission? Look for a hundred small things to improve. The results? We won that particular war.

Try Kaizen. You'll like it. Maybe going for the grand slam isn't for you. If the grand slam happens, great. In the meantime, make smaller wins your goal. One small improvement each day can add up to amazing results over time.

What small improvements could you make in your business starting right now?

11 Inspire Change: The Masters

Personal Or Professional Change: ADKAR Works!

ADKAR. The acronym is the invention of some change management gurus at a Colorado-based change management firm, Prosci. They've researched the topic of change carefully, skillfully blending academic and real-world perspectives. In addition, Prosci gathers lots of important data about organizations in the midst of managing change and reports that data to those who want to follow with change initiatives of their own. The acronym, ADKAR, presents change management to-dos in order, so read them that way.

A = Awareness of the need to change.

D = Desire to participate in and support the change.

K = Knowledge of what the change looks like and how to manage change overall.

A = Ability to make change happen day-to-day.

R = Reinforcement to keep the change going.

Want to know more? Order some Prosci materials. And read on too!

Are you achieving each aspect of ADKAR in your company?

YOUR NOTES:

12 Inspire Change: The Masters

The Studer Phases Of Change – Phase 1

Quint Studer is a well-known consultant in healthcare. He has developed his own Phases of Change. The idea is simple: before embarking on the journey of change, know what to expect, one phase at a time.

The Honeymoon

How do things look at start-up? There ought to be lots of excitement, along with some skepticism. Don't be surprised if there are plenty of questions in the minds and hearts of people. In addition, there is likely to be lots of hushed talk about possibilities as people try to cut the anxiety of uncertainty with speculation about what's about to happen. Questions without answers require answers, even if those answers are made up.

What to do during the Honeymoon phase of change? Well, for starters, provide ways to people to ask questions and get accurate answers. Make sure to discover and broadcast the wins resulting from change. Hold a quarterly forum to encourage dialogue, offering a way for questions to be heard and answers to be given. The more communication, the better. The pay-offs? To create a unified picture of the change as well as deepening commitment to it. Train leaders about the change itself as well as teach them how to manage the change. Those leaders can empower everyone else. And don't forget to communicate why the change is important in order to gain buy-in and support.

What phase of change is your organization or change project in?

How do you know?

13 Inspire Change: The Masters

The Studer Phases Of Change – Phase 2

Reality Sets In

Once the honeymoon is over, people begin to realize that accomplishing the change is going to be harder than was originally thought. Anxiety levels rise. Conflict surfaces along with the anxiety. People may start rationalizing mistakes and failures instead of honestly examining and learning from them. There may be more finger-pointing than usual as a climate of "us" versus "them" emerges.

To lead people through Phase Two, point out that this is just a phase and will not last forever. Explain the value of mistakes that are made honestly, not out of recklessness, negligence or carelessness. Encourage people to learn from failures and then move on. Offer more training about the change itself as well as how to manage it. More information means less anxiety, so pour it on. Keep broadcasting the wins. Listen to the tough situations. Answer the tough questions. Many experts talk about "over-communicating" throughout the change process, but particularly during this phase.

What phase of change is your organization or change project in?

How do you know?

YOUR NOTES:

14 Inspire Change: The Masters

The Studer Phases Of Change – Phase 3

The Gap Widens

Marathon runners often speak of The Wall. The Wall is the time during the race where the benefits of continuing don't seem to outweigh the pain. The pain is immediate, excruciating, palpable, and the desire to escape it is very real. The temptation to throw in the towel and quit is almost irresistible. Great runners know that this is a moment to of decision to "keep on keeping on."

In the same way, organizations going through change inevitably "hit the wall." Many will want to throw in the towel instead of forging ahead. Along with just plain wanting to quit, people may seem a bit discouraged, deflated and defeated. Even typically high performers may seem ready to relax their efforts at precisely the moment that ramping up is more important than ever.

During this phase of change, offer encouragement to offset discouragement. Train people to overcome resistance to change. Continue open communication, soliciting people's stories about the toughest challenges in order to provide not only tangible solutions but also an empathetic listening ear. Perhaps most importantly of all, begin coaching the low performers to get on board and improve performance or get off the train. Whatever you do, don't look the other way, hoping that low performers will somehow magically become enlightened, transforming themselves into high performers in the near future. It won't happen.

What phase of change is your organization or change project in?

How do you know?

15 Inspire Change: The Masters

The Studer Phases Of Change – Phase 4

Consistency

BREAKTHROUGH! If the change has been successful until now, much of the work done during Phase Three pays off. Everyone rallies as favorable results are experienced, understood, and celebrated. New practices become institutionalized, which is a fancy term for saying the new way becomes a customary way of life. What was once foreign, mysterious, and challenging becomes a habit.

Communication is clear and widespread. Customer satisfaction continues to rise. Employees are proud. The biggest danger during this phase? Complacency can set in. To combat that, keep things moving in a positive direction and set new goals.

What phase of change is your organization or change project in?

How do you know?

YOUR NOTES:

16 Inspire Change: The Masters

Kotter's Eight Steps to Organization Change – Step 1

John Kotter is a Change Guru consulting for organizations in change. Kotter has stated that at least 70% of organizational changes fail. To make sure your organizational change succeeds, follow Kotter's Eight Steps to Organizational Change.

Fire up the Troops

According to Kotter, firing up the troops, which requires creating a sense of urgency for change, is often skipped by companies, a potentially fatal mistake. Creating the sense of urgency to change sets the stage for every step to follow. While most organizations recognize the value of getting facts together to make a logical case for change, equally as many fail to make the emotional case. People's heads inform them about the nuts and bolts of change, including what needs to happen and how. But their hearts provide the spark that lights the fire of change, providing the heat to move and keep moving. So, how to fire up the troops?

Begin by taking the pulse of the organization to assess the people's sense of urgency. Are the people complacent, possessing an attitude that "Everything will be fine, as is." Or is there a prevailing attitude of false urgency, as people slog away to get the change accomplished but seem to lack the deeper spark or commitment to see change through? Or do the people seem genuinely urgent, inspired with a desire to win NOW?

Once you have taken the pulse of the organization, strengthen the sense of urgency to change—no matter what the current level—by appealing to people's hearts as much as their heads. Connect the vision for change to people's values, not just their logical minds, lifting their hearts as well as enlightening their minds. In addition, highlight the dangers of staying with the status quo and contrast that with the picture of change success.

How fired up are the troops in your organization for change?

17 Inspire Change: The Masters

Kotter's Eight Steps to Organization Change – Step 2

Create a Team to Guide Change

Get all leaders on board with the change. Create a Steering Team to specifically take on the change challenges. Choose a Change Sponsor from top leadership. Create the Steering Team and choose the change sponsor based on the formula: right people (skills & expertise) + right chemistry (credibility) + right power (position & leadership) + common goal (commitment to vision). Analyze the behaviors leaders must demonstrate for the change to have credibility. Ask every leader to live up to those behaviors. Install an accountability system connected to the change and hold people at all levels—from top to bottom—accountable for the change. Add the change and progress related to it to every regular meeting agenda. Based on meeting discussions, change progress, and leadership conversations, create "talking points" that every leader will cascade throughout the organization.

How effectively does out leadership lead change?

How do they prove they are walking the talk of this change?

YOUR NOTES:

18 Inspire Change: The Masters

Kotter's Eight Steps to Organization Change – Step 3

Craft a Compelling Change Vision

A clear vision simplifies what otherwise may seem complex and overwhelming. A clear vision also motivates people to act to make the vision a reality, even in the face of what might seem to be impossible odds. And, a clear vision creates teamwork, making it sensible and possible to coordinate decisions and actions that are aligned on a common purpose. The common purpose also lures people's attention away from day-to-day politics and on to something of far greater value—energy redirected to a cause with meaning. Great visions share 6 characteristics:

1. **Imaginable**: Easily seen in the mind's eye.
2. **Desirable**: Inspiring to stakeholders, even when mixed with skepticism.
3. **Feasible**: Comprised of realistic and attainable goals, even if those goals are very challenging.
4. **Focused**: Provide a basis to make day-to-day decisions.
5. **Flexible**: Clearly define the "what" but leave plenty of room for individual initiative in determining the "how".

6. **Communicable**: Easily and efficiently communicated to many at all levels.

What is the change vision for your project or organization's change?

19 Inspire Change: The Masters

Kotter's Eight Steps to Organization Change – Step 4

Communicate! Communicate! Communicate!

Remember 3 things about communicating through change, especially communicating the vision.

1. **Communicate about change, every day, every place, every level.** One channel of communication conveying one message and used one time is not nearly enough to make change understood, much less inspiring. Talk about change—especially about the vision and progress—in meetings, emails, documents, and publications. Communicate on computer screens and on signage. Talk about change formally and informally. Grab every opportunity to make the messages of change known loudly and clearly.

2. **Apply persuasive and informative communication skills**: Keep it simple**:** No techno babble or jargon.

   Make it vivid**:** A picture is worth a thousand words so use metaphor, analogy, stories and examples.

   Easy to pass on**:** Ideas should be easy to spread from person to person.

   Best in discussion**:** Dialogue is always better than directing.

3. **Actions speak louder than words.** Leaders who transform their organizations "walk the talk." They may be great verbal communicators, but they send even more potent messages by doing what they have said must be done.

Here are a few last tips about communicating through change. Be sincere and have conviction. Share the good news of change, the small wins and the large wins alike. Share the bad news of change as well, without holding back or watering down the message.

Be candid, but emphasize what can be learned from mistakes and setbacks. Solicit the views of others whenever you can and listen carefully and respectfully to objections, ideas, complaints and concerns.

Ask for questions, no matter how challenging, and answer them, even if you have to confess that "I don't know." Help others broaden their views, offering possibilities for different ways to think about things, even challenging long-held ideas, all with the aim of urging people not limit themselves to the party line.

How effectively is change communicated?

What are the methods being used?

What other communication methods could be used?

20 Inspire Change: The Masters

Kotter's Eight Steps to Organization Change – Step 5

Empower People to Execute

Execution is key to the successful accomplishment of change. All the best laid plans in the world and all the clearest communication possible won't mean much without action leading to results. But despite the fact that the need for execution during change seems obvious, organizations erect barriers to execution—often unwittingly. The barriers come in the forms of:

1. Internal processes, procedures and systems that contradict the aims of change
2. Supervisors and middle managers who don't want to change in order to properly…manage the change.

In order to manage the change process so that these obstacles don't block results, incorporate some preventive strategies.

- Install new processes and programs to support change performance, not inhibit it.
- Eliminate barriers to progress such as outdated processes and organizational impediments.
- Examine the impact of the company's structure on people's ability to achieve change goals.

- Make sure people are fully trained and coached about every aspect of change, technical and people.
- Clarify the leadership expectations for middle managers and supervisors about their roles during change.
- Provide training and coaching specifically aimed at helping middle managers and supervisors successfully make the transition.
- Ensure that middle managers and supervisors who don't fulfill new expectations are coached "up or out."

How effectively are organizational systems aligned with the change?

How effectively do leaders at all levels support the change?

21 Inspire Change: The Masters

Kotter's Eight Steps to Organization Change – Step 6

Celebrate the Small Wins

Research shows that companies that experience significant short-term wins by fourteen and twenty-six months after a change initiative begins are much more likely to complete the transformation.

The trouble is, sometimes the small wins don't happen as easily as might be liked. Add to that, when they do happen, someone has to notice in order to highlight and celebrate them. Create enthusiasm for the change and boost the momentum of positive forward action by:

1. **Discover the small wins.** Ask the Steering Committee for help with this as well as making sure leaders throughout the company have methods of sharing small wins. Hold them accountable for doing so.
2. **Highlight the small wins.** Publish them. Talk about them. Make them visible for all to see, in whatever fashion possible.
3. **Celebrate small wins.** People may have the natural tendency to downplay small achievements but you can't afford to do the same during change. Celebrate small wins.

Make small wins exciting; don't take them for granted. Start the round of applause and watch is spread.

How do you celebrate small wins?

Who among us can easily make that claim for our leadership? If we are wise, we will follow the lead of these magnificent examples if making change for good and change that will stick is our aim.

Perhaps the bottom line—and a key to it all—can be summed up in the idea of inspiration, versus dictatorship. While dictatorship carries costs too high to want to bear during change, inspiration bears other fruit entirely: commitment, clarity, initiative, innovation, collaboration, and risk-taking to name a few. The tree of inspiration seems the right one to be nourished by where change is concerned.

INSPIRING CHANGE

Great change is not dictated and, although high expectations are indeed required to make effective change, conveying high expectations for people's performance does not a dictator make. Dictators employ tactics quite different than high expectations. High expectations, when communicated expertly can cause any of us to reach for excellence. To the contrary, dictators demand blind obedience and submission—under threat of punishment and perhaps more—assuming no one has the talent to think or act sufficiently. All except the dictator, that is.

Dictatorship results in obedience and compliance at best and sabotage and rebellion at worst. It's tempting to rely on dictatorship—through change or anything else—as it seems expedient at first blush. It's not. It costs too much to be considered effective, efficient or anything so positive.

So if not dictatorship, what does work? The finest leaders of change in history have shown us the answers to that question. King, Gandhi, Lincoln, Kennedy, Roosevelt and leaders we know more up close and personally have inspired with the power of negotiation, storytelling, humor, metaphor, analogy and vision as their tools of choice.

They employed those tools well, as evidenced by the plain and simple fact that there are quotes from many of those inspirational change leaders that we all remember, along with their voices, facial expressions, hand gestures, stories and the results of all.

22 Inspire Change: The Masters

Kotter's Eight Steps to Organization Change – Step 7

Keep it Going

We just said that you need to celebrate small wins from early in the process of change and beyond. However, celebrating small wins does not imply "letting up" on the change process.

Keep the change momentum high. Persist with the efforts. Offer new challenges. Keep measuring and communicating progress. Continue over-communicating. Hold people accountable for progress. Continue to model strong leadership behavior. Walk around to hear about challenges and victories and spread the word throughout the organization.

In short, resist the all too human urge to start strong and assume the rest will take care of itself. It won't. As a leader of change, you are in charge of keeping things focused on the short-term and the long-term at the same time.

How do you keep the momentum for the change going?

YOUR NOTES:

23 Inspire Change: The Masters

Kotter's Eight Steps to Organization Change – Step 8

Make It Stick

I once visited an island that suffered from hurricane-like winds on one day of that visit. A full-fledged hurricane never developed—thank Heaven—but a few shrubs outside our temporary abode suffered nonetheless. They lacked roots deep enough to protect them from the daunting winds and so, blew over. In a very similar way, new practices must grow deep roots to be firmly planted in the culture or otherwise, may blow away in the wind.

Step 8 of Kotter's steps reminds us of that. Kotter insists that it's at this stage—at the end of the change process and not the beginning of it—that we must attend to culture shift. In order for the new way to become a true day-to-day "way of life," a real culture shift must take place. Here are Kotter's rules about culture shift:

- Cultural change comes last, not first, in the chronology of the change process.
- You must be able to prove that the new way is superior to the old.
- The successes of change must be visible and well communicated.

- You will lose some people in the process; some will not want to join the new culture; let them go with well wishes.
- You must reinforce new norms and values with incentives and rewards—including promotions.
- Reinforce the culture with every new employee, from day one.

In addition, Kotter reminds leaders to continue modeling the "new behavior." Align rewards and recognition with the new way. Tell stories reinforcing the new way, is the desired way. Use metrics to broadcast results from the new way, not the old way.

How are you making the change in your organization "sticky"?

24 Inspire Change: Reinforce Change

Say What You Mean…And Mean What You Say.

I once knew a partner in an accounting firm who took charge of an office. The office had been a poor performer within the larger firm. This partner was determined to turn it around? His strategy? Big doses of empowerment. He wanted staff to be more fully informed of all office activities. He also wanted to hear their voices more often about key issues. He also wanted teamwork.

So, he set the firm on the teamwork path. Along the way, the new executive realized that his plush corner office just didn't fit the teamwork culture vision. Add to that, the firm was growing fast and space was at a premium.

The new executive decided to change the firm's floor plans to include more open layouts. And his new office? Well, not an office at all, but a space right in the thick of things. He showed the team the finest team spirit.

How does your leadership reinforce the change?

YOUR NOTES:

25 Inspire Change: Questions

Ask "The Question"

Find out what the change really means. That means, ask questions. Lots of questions. When you do ask, use a tone of inquiry and curiosity, not a tone of interrogation. As Stephen Covey said, "Seek first to understand, before seeking to be understood." The idea is to open up communication, inviting candid responses.

A story I heard once illustrates this well. A VP of Marketing for a rather large company wanted to know what employees really thought about a new push to be customer focused. The annual company picnic was coming up, so the executive decided it was an opportunity.

The executive attended the company gathering...of course. She found herself sitting next to a young, smart employee, so naturally she wanted to know what the employee thought of the company's deliberate shift to customer focus. The employee happily shared her thoughts. The executive's biggest question?

What Has The Change To Customer-Focus Meant To You?

The executive heard many nice things. The employee was clearly very happy with the customer focus and had some very nice customer stories to share. After the executive left, the employee asked the person next to her. "Who was that?" The answer? "Oh, that was the VP of Marketing."

The employee breathed a sigh of relief. She was happy that she had shared positive news.

What questions do you ask employees about the change?

What answers do you get?

26 Inspire Change: Adjust

Line Things Up For Change

Adjust people, policies, and procedures to support change. List the new performance standards—those aspects of performance required for change to happen successfully. Then use the new performance standards to:

- Recognize people.
- Appraise performance.
- Promote from within.
- Hire and recruit from outside.
- Plan succession.

Line it all up for change strength. Fail to line it up and the contradictions may hurt change.

How well are our people systems aligned with the change?

YOUR NOTES:

ACTIVITIES FOR CHANGE

Effective change leaders must know the concepts and practices that make lasting and positive change possible. But knowing is not enough. Great change leaders must also be prepared to put what they know into action. They must be prepared to do things.

This book's last section focused on inspiring change, encompassing many ideas that leaders must have in mind throughout the change process—to make change successful and to make it long-lasting. This section will move from ideas to action, offering many tools that those invested in making change happen well can employ for success.

YOUR NOTES:

27 Activities for Change: Response to Change

Take The Team's Temperature

How is your team responding to change? Here is method for taking the team's pulse—quickly and easily.

A-Z, How Are Thee?

1. Create a "Temp Check" slip of paper for each team or group member. The paper should show the letters of the alphabet, printed in order. 'A' should marked by: "I am very angry and do not trust the company." 'Z' should be marked by: "We are in the midst of change and I trust that things will turn out okay." Note that there are 24 letters in-between.
2. Pass out the slips of paper, one per participant. Ask them to mark the place on the letters continuum that reflects their current attitude and pass in their slips.
3. Ask the group to guess where MOST answers from the quick survey will land.
4. Tally the responses and mark the totals on the continuum for all to see.
5. Address the issues: Is management really listening to the concerns of the organization? How could they listen

better? What would prove that they are listening (of course, you must LISTEN to the answers yourself)?

How well is management listening to the concerns of the troops?

28 Activities for Change: Resistance to Change

The Experience Of Change

Tune into how change feels and understand why people resist it. Try this activity as a way to remind people what change feels like and to start a discussion of how to handle resistance to change.

Change 3 Things

1. Ask participants to pair up. Partners should stand and face one another. Odd numbers? Make one group member an observer of the activity.
2. Give partners 20 seconds to look at each other and memorize the partner's appearance.
3. Ask partners to turn their backs to one another and change 3 things about their appearance. Even small changes count, such as removing a ring.
4. Ask partners to face one another again, looking one another over in order to discover the changes. Expect some fun and laughter this first time around.
5. Conduct a second round, using the same instructions as the first round. Expect a bit of confusion and grumbling. Make a mental note of it in order to point it out later.
6. Conduct a third round, using the same instructions as the first two rounds. Expect lots of grumbling. The grumbling,

of course, makes a key point: change is not always fun, especially if it seems constant. Before participants actually make any changes, give them permission to return to their seats.

How did you feel during each round?

What was the source of your frustration?

How does this activity remind you of changes at work?

What can you do to ease the pain of change and reduce resistance to change?

Recognize The Predictable Faces Of Resistance

The wise among us realize that resistance is normal and just one more thing to deal with during change. Want to ask employees about resistance? Use this activity.

Divide the team or group into smaller groups of 3-4 participants. Ask the small groups to discuss:

WHICH STAGE OF RESISTANCE IS STRONGEST IN THE ORGANIZATION NOW?

Choices:

Betrayal: "This cannot be happening here! Impossible!"

Denial: "This is just another management program. It will all blow over soon."

Identity Crisis: "Management has done this to us and they're stupid and greedy!"

Resolution: "We're on the right track now. This change was and is a good thing."

Ask groups to report back conclusions.

What stage of resistance is most pervasive right now?

What is the evidence?

What can we do about it?

Force Field Live

Gather the team and give this tool a try. Kurt Lewin invented the powerful change tool called Force Field Analysis. The idea is simple. Identify the Current State: How are things now? Identify the Ideal State: What is the best they could be?

Then, list all the Driving Forces: the forces that will push change toward the ideal. Also, list all the Restraining Forces: the forces holding change back. Here is an example of how it works:

Current State: Smoking 1 pack of cigarettes each day

Ideal State: No smoking

Driving Forces:

- Better breathing
- Nicer smells
- Saved money
- Better health
- Longer life

Restraining Forces:

- Cravings and withdrawal
- Increased anxiety
- Friends who smoke
- Stressful situations
- Loss of comfort

Lewin says that removing a Restraining Force or two is critical for making change successful. Ask the group to discuss:

What are the driving forces and restraining forces of the current change?

What restraining forces can we remove or reduce?

31 Activities for Change: Trial By Jury

There Are Two Sides To Everything

Learn about Driving and Restraining Forces during change. Help people work through their resistance during change by asking them to play Trial By Jury. It's not just enlightening and a way to encourage discussion; it's fun!

Trial By Jury

Gather a group of change stakeholders, such as employees. Ask the group to create arguments for and against the current change. Then, let them pronounce final judgment. You'll likely learn plenty about the Driving Forces and Restraining Forces of the current organizational change.

1. Divide the larger team or group into sub-groups. Prosecution. Defense. Judge/Jury.
2. Explain the roles. Prosecution will argue for why the Current Change is wrong. Defense will argue for why the Current Change is right. Judge/Jury will make the Final Judgment after all arguments have been heard and one round of counter-arguments have been made.
3. Give the Prosecution, Defense and Judge/Jury 15 minutes to prepare arguments.
4. Ask for presentations. Prosecution should go first. Then Defense. Then Counter-arguments.

5. Ask the Judge/Jury team to pronounce the Judgment and Sentence (prescribe actions to take).

6. Review some key points and observations from the whole activity.

What are the best arguments for change?

What are the best arguments against change?

What can we do to strengthen the forces for change and reduce the forces against change?

32 Activities for Change: Have The Best..."

Look At The Bright Side

Make it a game to see the positive. Here is another activity from Benjamin and Rosamund Stone Zander as shared in *The Art of Possibility*.

Play a game of "Have the best ___________ever." Play the game regularly throughout the change rollout. The aim of the game? To boost spirits by directing people's attention to pleasant activities and outcomes.

Here's how the game works. Fill in the sentence blank with any everyday activity that the team can focus positive efforts toward. By doing so, you will steer people's attention to what they can control, instead of what they cannot control. Make the activity playful and inspiring.

Review the game results at the end of the day or week that the game is played. Some possible assignments might be:

- Have the best CUSTOMER EXPERIENCE ever.
- Have the best BREAK ever.
- Have the best TEAM EXPERIENCE ever.

- Have the best CONVERSATION ever.
- Have the best DAY ever.

What were the best experiences?

What does the activity teach about what can be controlled and what cannot be controlled?

33 Activities for Change: Even The Best Lose

Change Gains And Losses

The gains of change may be clear. The losses are better faced, not dismissed. Help people understand the WHOLE picture of change, gains, losses and all.

Even The Best Lose

1. Ask participants to list some positive changes they have experienced. List the changes on a flip chart. Examples: Promotion, Marriage, Birth of a Baby.
2. Pick a change. Investigate the gains and losses by asking the questions: What were the gains? What were the losses? Chart some answers.
3. Remind participants that change—even change that is chosen and eagerly anticipated—brings an accompanying sense of loss. It's not right or wrong. It's just inevitable.

How can we honor the sense of excitement of change?

How can we honor the loss experienced during change?

YOUR NOTES:

34 Activities for Change: Status Quo Busting

Challenge The Status Quo

Get people to change...on purpose. People settle into comfort and predictability. That's part of what makes change so hard.

Train people to look past comfort and predictability and strive for changes of their own making. Use this Status Quo Busting activity with your team. Do it often. Practice makes perfect.

Gather the team. Explain that constant change is a way of life and that we all need to take charge of change before it takes charge of us. Pose the question:

> WHAT ARE ALL THE THINGS WE DO THAT FALL UNDER THE HEADING, "THAT'S THE WAY WE'VE ALWAYS DONE IT?"
>
> Get everyone on the team to contribute items. Make a long list. Then, review the list. Ask:

What 3 items should we tackle first?

Where would we get the biggest gains from improving the way we do things?

Discuss all the items on the list. Evaluate the alternatives. Decide on 3 items to improve. Create improvement action plans and assignments. Set follow-up and review dates. Evaluate progress and celebrate positive results.

35 Activities for Change: Group Mood Ring

Do A Mood Check

Check the mood of your team regularly during change. This activity applies the principle: If people share their feelings, they can manage those feelings.

Use A Group Mood Ring

How to conduct a Group Mood Ring? Here's how.

1. Distribute 3 blank index cards to each team member. Ask participants to choose 2 moods that best represent how each is feeling about the change right now, at this moment. Ask them to write those moods on 2 cards, i.e. excited, sad, cautious, tired, exhausted, angry.
2. Ask participants to consider what mood they would like to achieve related to the change and ask each to write that mood on 1 card.
3. Ask participants to pair up in order to share their current and desired moods. Ask pairs to discuss the question: How to achieve the desired mood?
4. Conduct several rounds of the activity, asking participants to pair up differently each time, if desired (this make a nice mixer).

5. Poll the larger group for moods and methods for changing those moods for the positive.

What moods are you experiencing now?

What moods would you like to achieve?

How do you intend to do that?

36 Activities for Change: Puzzle Pieces

Pieces Of The Whole

Bring the pieces of a puzzle together to make change work for you, not against you. Here's a really fun and invigorating way to prove the point: Each team member holds a piece of the puzzle that can change the view of the whole.

1. Find some pictures of famous, well-known paintings. Reproduce the photos, making sure each is large enough for a puzzle. Cut each photo into 6 pieces. Each piece ought to be approximately the same size and shape for ease of assembly. For example, simple squares will do. You will need one puzzle for each group of 6 participants.
2. Mix the puzzle pieces and distribute them randomly throughout the larger group. Hold some extra pieces so you can share them with the right groups at puzzle assembly time.
3. Ask participants to give a descriptive name to their individual pieces of the puzzle.
4. Ask participants to gather with those who hold companion pieces to their puzzle piece—their small group.
5. Once participants are with their small group, ask them to assemble the entire puzzle and give it a descriptive name.

6. Review the names with all participants. Show the completed puzzles—the whole view—via PowerPoint or similar format, along with their famous names.

What did you name your individual piece of the puzzle?

What descriptive name did your group give the whole puzzle?

How did the names differ?

During change, how might an individual perspective differ from the view of the whole?

37 Activities for Change: Resiliency

Reflections On Resiliency

Help people become tuned into resiliency concepts and practices. Conduct this Resiliency Activity with your team or group. See what discussion about resiliency you can stir up.

Hanging In There For The Long Haul

1. Ask participants to divide a blank sheet of paper into 4 quadrants.
2. Tell participants that they should draw a simple figure in each quadrant and that they'll have 15 seconds to complete each drawing. Explain that the drawings will begin in the upper right quadrant and move clockwise around the page.
3. Ask participants to draw a house in the upper right-hand quadrant. Track the 15 seconds of time from start to finish.
4. Ask participants to move to the lower right-hand quadrant and using their non-dominant hands, each should draw a cat. Again, track time for 15 seconds.
5. Ask participant to draw a flower using their dominant hands, but with eyes closed. Again, allow 15 seconds for drawing.

6. Final drawing. Ask participants to use their non-dominant hands and close their eyes as they each draw a tree.

As you completed each drawing, how did your attitude change?

What did you think?

How did you feel?

What does this activity remind you about change?

38 Activities for Change: String Thing

Push Or Pull?

A key ingredient to successful change is illustrated by this simple activity. To remind people about the Vision Of Change, use this activity.

A String Thing

1. Give each team or group member a piece of string, approximately 12 inches long.
2. Ask each participant to vertically place the string in front of him or her. A hard surface works best.
3. Instruct participants to place an index finger on the end of the string closest to him or her and PUSH. The string should bunch up and not move very far forward.
4. Next, participants should place their index fingers over the end of the string farthest away from him or her and PULL. The string should move forward.

How did the motions of push and pull produce different results?

How does the activity relate to change?

What is our current "vision of change" and how effectively is it pulling us forward?

What did you think?

How did you feel?

What does this activity remind you about change?

39 Activities for Change: Potato Game

Lessons From A Potato! Yes!

Capitalizing on change requires practice, momentum, and follow-through. Want to teach the troops some change lessons? Try this exercise in learning by doing. Please practice and master the process before asking others to give it a try.

The Potato Game

1. Provide 1 potato for each team or group member. Choose potatoes that are solid but not too hard. Russets work well. Also choose potatoes that are large enough for a participant to easily hold the potato and strike it with a straw.
2. Provide 1 drinking straw for each team or group member. Choose plastic straws that do not have any bendable portions and are stiff.
3. Demonstrate the activity. Hold a potato in your non-dominant hand. Grasp the potato, fingers forming a "C", around the potato. The hand must be protected so it is not accidentally pierced by the straw.

   Grasp the straw with the dominant hand, thumb over the end of the straw that is opposite from the strike end. Begin moving the dominant arm up and down from the elbow through hand, continuing to hold the straw, thumb over the end. Get a solid and smooth momentum going.

When ready, place the potato in the path of the downward motion of the straw/hand/arm. The straw should easily pierce the potato all the way through.

4. Ask participants to follow your lead, providing coaching for those who want it.

SAFETY TIP: Keep all body parts away from the underside of the potato to prevent any straw stabbing.

VARIATION: Ask for several volunteers to go first. Ask them to coach the rest of the group. Note the rapid progress through the group.

LESSONS: Practice produces successful performance. Coaching helps. Momentum is important during change. All good change requires follow-through (the straw only goes through the whole potato via smooth, long strokes, not quick jabs and stabs). Those who successfully go first should pass it on.

What are the lessons from potato stabbing?

How can the lessons be applied to our change?

40 Activities for Change: Magic Wand

Make Three Wishes

Get people used to the idea of change by asking them to talk about it.

The Magic Wand

1. Explain to the group or team that this activity will help them consider change and each other in a new light.
2. Give these instructions. "You have just found a magic wand that allows you to change three work related activities or aspects. You can change anything you want. How would you change yourself, your job, your boss, co-workers, an important project, etc.?"
3. Direct discussion once around the group or team to hear the answers.
4. Lead a discussion about how and why participants selected their changes.

This activity will bring more ideas for change to the surface. In addition, the group or team will become more accustomed to the idea of constant change. And, the group or team will get to know each other's interests and values at work.

What changes would you make and why?

What changes will lead to the most lasting, long-term positive results?

41 Activities for Change: What I Think...

Finish The Sentences

Discover what stakeholders really think about change.

What I Think Is...

1. Gather the group or team together. Or do this activity at a regularly scheduled meeting.
2. Explain that the purpose of the activity is to discuss views about change.
3. Use a flip chart or projector to show the group or team some sentence stems.
4. Give the group or team a few minutes to write answers to each stem.
5. Ask each person to share his or her answers.
6. Facilitate a discussion of what can be learned from the exercise.

Some sentence stem examples:

- When I think about this change, my first thought is...
- What I like about change is...
- What I hate about change is...
- The first thing that comes to mind when I think about change is...

- The most important thing for this team to keep in mind about change is…
- We could ensure the success of this change by…
- The biggest obstacle we will face during this change is…

What can we learn about change from hearing people's answers?

How can we use the answers to manage change effectively, as individuals and as a team?

42 Activities for Change: Card Game

It's In The Cards

Want ideas from everyone on the team? Use this activity to gather many ideas quickly and get people talking.

A Card Game For A Change

1. Give each member of the group or team 4 blank index cards.
2. Create a question about change that you would like each team member to answer.
3. Some question examples:
    a. What are the supports and obstacles we face during this change?
    b. What do we need to do to make this change a success?
    c. What are your questions about this change?
4. Pose the question to the group or team and ask that each person write 1 answer per card.
    a. Suggest that more detailed answers work better than broad-brushed ones. Offer an example.
    b. A variation: If desired, pose 4 questions and ask that an answer for each be written on the 4 cards. In that

case, make sure that cards are also numbered in sync with the questions.

5. Collect all the cards and shuffle them. Then, deal the cards again, 4 per person.
6. Ask the group or team to mix for 5-7 minutes, sharing their cards with as many other participants as possible.
7. Ask participants to return to their seats.
8. Discuss any interesting ideas shared during the mixer.

What can the answers reveal about change?

How should we use the answers to improve change success?

43 Activities for Change: Rearrange

A Change Experience

Start a discussion about change by providing a change experience.

Rearrangements

1. Before the group or team gathers, change the usual seating arrangement. Make sure the participants will be facing the same way and toward what is usually the front of the room. Also, place the chairs in straight rows, classroom style.
2. Set your stage BEHIND the participants.
3. Start the presentation by sharing the following. "Change often happens in the same way that this room is arranged. The leader gets behind the employees and pushes. The chances for success when change is approached this way, is about the same as our success if we conducted class this way.

   In addition, many organizations are structured like our neat line-up of chairs. Straight. Rigid. Things are hard to see. It's often equally hard to communicate. And hard to learn from each other. In a real team, this should not be the case, not if we want to succeed. We must get out of the box for change to succeed."

4. Ask the participants to rearrange the room for top teamwork, the best communication and effective change.

How can we arrange things—right down to seating—for effective change?

44 Activities for Change: Poker

Get 'Em Involved!

Poker Anyone?

Play poker to get people involved. At the beginning of your meeting or class, explain the rules of the activity.

"In order to encourage involvement during this meeting, I am going to pass out a playing card every time someone speaks up. Your objective is to gather as many cards as you can. At the end of our meeting, the person who has collected cards to make the best poker hand will win a prize."

Here are a few tips to make the activity successful. Agree on the poker rules in advance. Ensure that you have plenty of cards to cover size and level of potential participation in the group. You can and should use more than one deck depending on the number of participants. Have enough prizes in case of a tie.

A variation? Use the card reward for other behavior. Give cards out for those who return on time from breaks or lunch. Or offer cards to each participant who asks an excellent question. Or mix it up by changing the conditions for winning a card throughout the meeting.

What are the best ways to encourage full participation during meetings?

YOUR NOTES:

45 Activities for Change: Picture It!

Vision Review

Picture It!

Answer the question, "How does the vision of this change affect you?" Discover how the group or team believes the change will affect them. Do this Vision Review.

1. Explain the purpose of the activity: to share how the change vision affects participants.
2. Ask the participants to close their eyes, lay their heads down on the table, or do whatever is needed to get comfortable for the activity.
3. Share these directions, speaking slowly, calmly and clearly. "In order to do this Vision Review, please keep your eyes closed and stay as comfortable as possible. As I share the vision for the change from the organization's perspective, please picture how you think the change will affect you in your day-to-day activities.

   What will you work on? How will the job be better? Who will you work with? How will your work area look? What will be discussed at meetings? How will customers react differently? Picture every aspect possible. We'll spend at least 5 minutes in our vision trip. Let's begin.
4. Conduct the activity. Turn down the lights to create a calm and reflective environment. Use a slower and calmer pace

for talking than usual. Repeat the questions more than one time, spacing them out over several minutes. Give participants plenty of time to visualize answers.

5. After 5 minutes, ask participants to open their eyes. Ask them to share aspects of their visions.

What are the effects of the vision of change on each individual in the organization?

46 Activities for Change: Point Out Obstacles

Visualize This

Point Out The Obstacles

Metaphors make a great point and sometimes several great points. Help the group or team understand the value of deliberately looking for change obstacles with this metaphor.

Imagine the gutters on your house. It's storming and the gutters are directing the torrents of rain to safer ground. What could block the gutter's effectiveness? Leaves. Fir needles. Pine cones. Dirt. An accumulation of these pose slow flow and may even stop the work of the gutters entirely.

What if the gutters work inefficiently or stop working? What happens? Basements can flood. Gardens can drown in a puddle of water. Foundations can crack. Roof rot can set in.

Now, imagine the change our organization is undertaking. We want success. We want the change to flow freely throughout the organization, supporting progress toward fulfillment of the change vision. It is likely, however, that we will meet some obstacles along the way. What are they? Imagine the obstacles to change. See them in your mind's eye. We'll share thoughts about those obstacles in a few moments.

After a few moments of visualizing, ask for descriptions of obstacles. List them on a flipchart. Conduct some further discussion.

What obstacles do we envision?

How can we prevent or overcome those obstacles?

47 Activities for Change: Line Up Game

The Line Up Game

Conduct this physical and visual change assessment.

1. Ask the group or team to line up. Random order.
2. Explain that one end of the line is "Agree." The other end is "Disagree." Point out the "Agree" end of the line versus the "Disagree" end.
3. Pose several "Agree" or "Disagree" statements about change and ask people to physically choose a position in line that best represents their level of agreement or disagreement. People may group if several choose a similar spot on the continuum; no need for a perfectly straight line.
4. Once people are positioned (after each statement is shared), ask participants to discuss with neighbors why they chose their positions. They should discuss with those near them on the line, 3-5 will do.
5. Then, ask for a few people along various points of the continuum to share with the large group why they chose their spot.
6. Repeat the process for as many questions about change as you wish. Some question examples.
    - The current change is going very well. It's going smoothly and with no hitches.

- I love change and can't wait for the next change to happen.
- Organizations should change often to keep up with the competition.
- Change is inevitable in this hard-charging, competitive world.
- Management is doing a good job with this change.

Here is one piece of advice. Don't be shy about hearing bad news. If people are honest, you'll get some. Of course, the wise among us know, that's exactly what you want. Good news is fine, but it's not the only worthy news.

How do people view change?

How effective is the organization's current change?

How well is management managing change?

48 Activities for Change: PIN

A Discussion Management Technique

Get a variety of answers from people using this strategy.

Positive, Interesting, Negative

Keep the group or team discussion balanced by using PIN. The acronym stands for:

P = POSITIVE

I = INTERESTING

N = NEGATIVE

Ask participants to prepare comments on a particular topic by using PIN. Each must share a positive, interesting, and negative opinion related to the topic. Do the activity round-robin style and you'll have balanced participation with balanced views.

What is positive about this change?

What is interesting?

What is negative?

YOUR NOTES:

49 Activities for Change: KSS

Check The Change Process

KSS

Use this simple assessment to gauge the progress of change. A simple way to check the group or team's views of how things are going is KSS. The acronym stands for:

K = KEEP doing this.

S = START doing this.

S = STOP doing this.

Here's how KSS works. Ask the team to appraise the progress of the change (or anything else, for that matter) by asking each to express a Keep, Start, and Stop. Record the answers on a flipchart. Discuss more in depth, especially where you notice patterns in answers.

What should we keep doing?

What should we start doing?

What should we stop doing?

YOUR NOTES:

50 Activities for Change: Bank Shot

Question Bank

Getting people to ask questions is very important.

A Bank Shot

One of the most powerful ways to manage resistance to change is to answer the questions people may ask. Or to answer questions people are thinking but are afraid to ask. Make a "Question Bank" so people have an easy and anonymous, way to pose tough change questions. How to do it?

In a common area, place a box or bank with a slit in the top. Ask people to put their current questions about the change in the Question Bank. Tell them you will answer randomly selected questions regularly (best to be specific about the dates, times, and places).

Let's say you pick the weekly staff meeting, for example. At staff meeting, select 1 or 2 questions—or more—and answer them. Give accurate, thorough and specific answers. In cases where you don't know the answer to a question, simply say so. "I don't know." is the only proper answer…if it's true.

What questions do stakeholders have about the change?

How can we hear and address those questions to positively deal with resistance?

What are the best answers to give?

What should we do if we "don't know" an answer to a question?

REFRAME MYTHS

The Myths and Realities of Change come to us from change guru, Price Prichett. His point? To lead change effectively, one must understand the typical human responses to change: the visceral reactions that many people experience when the challenges of change arise.

The thing is, just because an experience feels so convincingly real, does not mean it reflects truth. In order to be effective in life and work and during change, it's important to distinguish what's true—founded in fact—and what's not. This section will outline those myths and realities. Use them to enlighten yourself as well as to help light the path forward for others during change.

YOUR NOTES:

51 Reframe Myths: Organizations and People Suffer...

Because of the Myths of Change.

People have some understandable—but not very useful—ideas about change. Let's call those ideas the Myths of Change. The Myths of Change are myths because—despite the fact that they are commonly held—the myths simply are not true.

Add to that, unlike the mythical folklore of ancient times, the Myths of Change more often create pain and suffering, not hope and possibility. As such, the Myths of Change themselves are well worth...changing. What to put in their place? Some doses of reality.

To discover whether you endorse more Myths of Change rather than change realities, read on for several pages. Read each Myth of Change. Read the contrasting realities. Ask yourself, "Am I a myth holder or a reality seeker?" Believing in the Myths of Change will frustrate you and not much else.

By contrast, the realities of change can empower you to ask more, prepare more, learn more, and ultimately become more, as both an employee and a person. Go for reality.

Think about the myths in your organization:

Which of the myths do you endorse today?

Which realities would empower you for tomorrow?

52 Reframe Myths: Myth #1

Myth: The problems we've had because of this change prove that the change is bad.

Reality: While on the one hand, change initiatives are designed to ultimately solve problems, on the other hand, change creates problems as well. How can this be? Well, for starters, change requires experimentation and risk. New methods must be employed. New tools will be used. New solutions will be put into action.

People with so much new to do have so much new to learn as well. Mistakes and the problems that go with mistakes go hand in hand with that learning.

A robust change plan will prevent some but not all of those mistakes and problems. But what about the problems that were unforeseen? Chalk those problems up to the price of progress. Our best bet is to simply learn every lesson there is to learn from those unplanned bumps in the road.

Yes, there are problems during change. It's true. But do the problems of change make it right to assume the change itself is bad? No. Let's not throw the baby out with the bath water.

How can the change's problems offer opportunities?

What can we learn from the inevitable problems of change?

53 Reframe Myths: Myth #2

Myth: Top Management—"they"—are hiding something.

Reality: The truth is, Top Management is likely sharing all that they know today. In instances where the messages from yesterday sound different than those we are hearing now, or in instances where there are no messages at all, Top Management has not necessarily lied.

Instead, it is more likely that today's information is more accurate than yesterday's or that there is no updated information at all and the messages coming from the management suites reflect that.

The bottom line is, in most cases, when Top Management knows more, they'll tell more. What looks like hiding is simply Top Management thinking things through. Or, it may be Top Management waiting for new and more reliable information before making new pronouncements.

What are the reasons that management may not be sharing answers about change?

YOUR NOTES:

54 Reframe Myths: Myth #3

Myth: If I get mad, I can fix things. I deserve to be angry because things are screwed up. Things SHOULD go smoothly—and be more in control. They aren't, so I am mad.

Reality: When reality drags at the heels of high expectations, people get angry. It's understandable. If the anger resulting from the reality versus expectations gap served to fuel constructive problem solving to close the gap, the anger might even be productive.

Unfortunately, in the case of change, that positive use of anger too often does not happen. Instead, the anger just fuels more anger and a general restless and discontented mood among the troops that does not raise performance, but undermines it.

So, okay, get mad when reality lags high expectations. But, don't stay mad for long, even if others choose to get stuck there. And don't tell yourself that you "deserve" to be angry because you deserve so much more than that.

What you really deserve is the sense of empowerment that comes from solving problems, not just dwelling on them. What you really deserve is to focus on what you can control during change, not on what is outside your reach. What you deserve is to learn and to grow and to succeed, even if it requires many trials, and venturing into arenas that you at first resist and where others dare never to go, but only want to complain about.

The bottom line? What you deserve are the benefits of change, but you will only enjoy those, when you give up your righteous indignation over the mistakes, snags and other issues related to it.

What are the expectations of change?

What is the reality of current performance?

What is the change gap?

What can we do about it?

55 Reframe Myths: Myth #4

Myth: Management just doesn't care about how the change is going. Or about how it affects us.

Reality: While it may appear that management does not care about the change or its effect on all employees, it's much more likely that management cares in ways we'll never know. Yes, it's true that perhaps it would be better if management's care was much more obvious. But, don't mistake what appears to be lack of care for the real thing, unless you can verify that as a fact.

While some management people may not care personally for each and every one of us, you can be reasonably sure they want the company to succeed.

And many of those same managers care very much about how people are affected by change. Likely most do not take the impact of change on people's lives and work lightly. In the end, the majority of managers are very interested in doing what is best for the most people, most of the time.

How do we rate the level of caring of our managers during change?

What is the evidence?

YOUR NOTES:

56 Reframe Myths: Myth #5

Myth #5: Management is clueless and incompetent.

Reality: It's more likely the case that management has put some real thought into the change. They have likely discussed the change in depth, even if behind closed doors. It's highly probable that management has made a business case for why the change should happen and out of that, crafted a change management plan. In the end, far from being clueless and incompetent, management has likely put a fair amount of work into the change.

At the same time, management people are human. They make mistakes too, right along with the things well done. They are learning too, just like the rest of us. They make plans, some parts of which go well and some that flounder. At any time during the change game, a scorecard applied to management performance would likely show plenty of hits, run, and errors.

Competence-Incompetence is properly measured on a continuum, not in black and white, binary fashion. Looked at in that light, it would rarely be fair to accuse any management team of being clueless and incompetent.

What does the management team do well?

What could the management team improve?

YOUR NOTES:

57 Reframe Myths: Myth #6

Myth: If we wait long enough—just doing business as usual—this change will simply die a natural death and we will never have to deal with it at all.

Reality: Waiting until the current change blows over may seem like a good idea. Sometimes that strategy may even work as some changes do fail. They fizzle. They sputter. Some do die a natural death.

But, there are some flaws in the ointment of this Myth of Change. First off, by avoiding positive involvement with the change—even if it eventually fails (which it may not)—you lose an opportunity to stretch and grow. Sure, the status quo can be comfortable, but it's rarely exciting. Change, on the hand, is stimulating, energizing, and enlivening. Why miss it?

Secondly, there's danger in sitting around imagining that the change will just "go away". What if it doesn't just "go away"? By the time you get around to boarding the train, you may have a hard time catching up with it. Is that really a risk you're ready to run?

Take change on now and at least learn some things. Wait and potentially pay the consequence as you huff and puff to catch up with the rest of those already on the train.

How likely is this change to just "go away"?

What are the advantages to getting on board with change now, not later?

58 Reframe Myths: Myth #7

Myth: I'll ride this out and do my job just as I have always done.

Reality: Watching and waiting can seem safe and secure, but despite outward appearances, this strategy has its hazards.

While appearing to be quite a challenge, the change presents an opportunity to upgrade knowledge, skills, and overall performance on an individual and an organizational level. Every day that you watch and wait means that opportunity is passing by.

In addition, as the opportunity to sharpen knowledge and skills passes, so does the chance to notch up excellence. Surprisingly and sadly, you may move one notch closer to mediocrity instead.

So, don't just watch and wait, doing your job today and tomorrow in the same way you did yesterday. Challenge yourself! Change affords you that chance. Take it and grow with it! You may even thrive along with it, no matter how the change itself turns out.

What knowledge and skills might you acquire during this change?

How can you reach for excellence on purpose?

YOUR NOTES:

59 Reframe Myths: Myth #8

Myth: What can I do? I don't matter! My efforts are just too small and too little to count.

Reality: You DO matter. You DO count! Every action you take, every single day, has a ripple effect, far reaching and with consequences you cannot track and therefore, may never know.

Whether you can see visible and immediate results of what you do or not, be assured there are results. Assume that and guide your efforts based on that assumption.

What does the belief that you do count mean during change? It means that every time you ask a question, not only do you become more informed, so do others. It means that every challenge you tackle, even while others look on cynically, demonstrates that the way forward is better than sitting still. It means that your positive actions count, even when all those around you may think, talk and act to the contrary.

No, we don't always see all of the positive results of our efforts right away. In fact, some positive results may never be seen by our own eyes at all. But, our efforts—for better or for worse—count for far more than we could ever imagine. Don't sell yourself short or small, because the plain and simple truth is that you're not.

What individual actions can support this change effort, even in a seemingly small way?

60 Reframe Myths: Myth #9

Myth: Changes? We don't need to change! We are fine the way we are. We are successful. We are good. Why change? Saying that we need to change, is the same as saying we are sub-par. That's unfair!

Reality: We live in an age of intense global competition, significant economic challenge, blistering technological change, and immense cultural diversity, just to name a few factors urging those who wish to continue to be successful in the future on.

Under the conditions of our times, the wise know that they dare not stand still. The wise know they need to change BEFORE there are serious problems, disruptive competitors, compromised supply chains, and on and on. Not after.

Change is part of a full life and a successful business. In either case, if we stand still, we become second-rate, perhaps even obsolete and ultimately we die.

What are the external factors propelling this change forward?

What are the dangers of standing still?

YOUR NOTES:

61 Reframe Myths: Myth #10

Myth: What about my job? Am I safe? This change looks fishy to me. I'll bet that while others have plenty to gain, I have plenty to lose.

Reality: Yes, you have plenty to lose. You can lose your current level of comfort. You can lose the sense of security of knowing exactly what to do every minute of every day. You may lose a sense of power over information that only you held yesterday as it all gets freely shared during this change.

You could lose a sense of control over methods you knew inside and out as you transition to new ways of doing things. You can lose the sense of failure over making mistakes. There is plenty to lose and much of it is quite emotional, whether you ever talk about it in those terms or not.

Here is the flip side of the change coin. There is plenty to gain too. You can gain excitement and stimulation as you learn more than you have in years. You can gain the satisfaction of sharing power as watching others succeed has its own benefits.

You can gain the curiosity of exploring new ways of doing things. You can gain a sense of freedom from learning from mistakes instead of fearing them. The list of potential gains could go on and on.

Sure, you can lose a lot during change. But for everything you could lose, you could just as easily site and state a potential gain. Perspective is everything during change. Cultivate yours to work for you and not against you.

What losses could happen during this change?

What gains could result from this change?

62 Reframe Myths: Myth #11

Myth: "They" should be on top of these changes. It's not my job to deal with this.

Reality: Let's begin by asking, "Who are 'They'"? Claiming that there is a nebulous "they" who ought to be in charge of this change implies that you can handily point the finger. Cast blame. Pin the tail on someone else's donkey. And further, you have nothing to do with either the success or failure of the change as that obviously happens "somewhere else".

Really? We think not. In fact, the phenomenon of "they did it" is quite impossible as the success or failure of anything of importance in your company is the responsibility of... EVERYONE!

So, take the following pledge. Repeat to yourself. This is MY company. I choose to support it and build it, day by day, via my performance. Or not. I can be part of what is possible along with everyone else. Or not. I can be part of the problem or part of the solution. The choice is mine. Or I guess I can also choose "not."

When I sit back and point fingers or pass the buck, I disempower myself and may even convince myself of the lie that says "They are guilty and I am not." But worse, I also lose the chance to boost my confidence and spirit by being a positive contributor.

What can you do to contribute positively to the change?

ACCOUNTABILITY FOR CHANGE

"No one can make you feel inferior without your consent."

~Eleanor Roosevelt

Eleanor Roosevelt understood accountability. She knew that in order to live full, satisfying lives we must take accountability for ourselves right down to our very thoughts, and certainly, to include our actions. This is hard, to say the least, but anything less poses roadblocks to our fullest growth and development, especially during change.

This section provides thoughts, tools and methods for appraising one's level of accountability during change. Awareness is key. Taking action for the better is also required. Read on and find ways to live up to your highest accountability and along with it, realize your finest potential in life and in change.

YOUR NOTES:

63 Accountability: Danger Or Opportunity?

How Do You See It?

The Chinese symbols for change convey two meanings: "danger" and "opportunity." The lesson is clear. Change brings both risks and rewards. The question is, "When you think about change, what comes to mind?" Risk? Or reward?

If you're like a lot of people, risk comes to mind first. And perhaps dominates. The upside of change is harder to picture. And that picture of reward is even harder to keep.

When it comes to change, do you think more about risk or reward?

Consider a change you are part of. What are some of the risks and rewards?

YOUR NOTES:

64 Accountability: ABT Or DBT?

You Choose!

ABT? DBT? What are we talking about? Katherine Cramer and Hank Wasiak tell us all about Asset-Based Thinking and Deficit-Based Thinking in *Change the Way You See Everything.*

Asset-Based Thinking

- What is right and good?
- Where are the opportunities?
- What are my strengths?
- What are his/her strengths?
- What CAN we do?

Deficit-Based Thinking

- What is wrong and bad?
- What are the dangers?
- What are my weaknesses?
- What are his/her weaknesses?
- What is holding us back?

While our nervous systems naturally sensitize us to problems and danger—our survival depends on it—a steady diet of "being on alert" wears us out. It curbs our creativity, our zest, and our enthusiasm.

Add to that, many of the things we perceive to be dangerous—in this modern world, at least—are inconveniences at best. They are not truly dangerous at all. And, since that is the case, all of them can be viewed from a more positive angle.

For example, during change, many people fear the loss of control of having to shift to new methods. What happens when one has to learn a new computer program as just one example?

Fear of failure, because there will be some. We question ourselves. What if I make mistakes? What if I am not as efficient?

The deficit-based thinking is natural as the fears are real. There will be mistakes and there will be less efficiency. But while our bodies register such fears in the same way as we would if a saber-toothed tiger from ages past faced us, the dangers are not comparable at all. Mistakes do not lead to death but can lead to quite the opposite: growth and enlivenment.

What are some fears faced during change?

How could those fears be transformed into opportunities?

65 Accountability: Think, Think, Think!

Change your thoughts...Change your work.

When our thinking is out of whack, it pays to know it. But many people do not. Such people go about their lives simply thinking and saying things that are half-truths at best, all the while pretending that the inaccuracies are facts. Actually, it's probably safe to assume that we all do it from time to time.

What difference does this make? Whacky thinking arouses emotions not in line with the reality of the circumstances. For example, we react to a question as if it's a challenge to our competence—and we're tempted to flare up.

This is a case of our survival system functioning in a situation that does not demand life or death reactions. The result? Emotions surge and rational thinking shuts down. This is not a good formula for solving today's organizational problems. What's the antidote? Dispute the irrational. Something like this:

You Hear...	You Respond...
"I can't stand this!"	"Oh really? But you already have… stood it, that is."
"This is the worst!"	"The worst? You mean, nothing could be worse? Not likely."

"I'll never learn this."	"Never? Yes, it might be hard and take some practice, but certainly, you can get it."
"Nobody cares but me."	"Nobody? Are you sure? There must be others who care."
"This is terrible!"	"It may be a disruption, or inconvenient, but terrible? Not likely."
"This is a waste of time."	"Isn't there possibly some good reason to do it?"
"It's all about money."	"Is money the only reason?"

What irrational thoughts and statements could you challenge?

How would you challenge those thoughts?

66 Accountability: The Stress Of Change

How Do You Handle The Stress Of Change?

Are you Type A Or Type B?

And how do the Type A and Type B personalities handle change? Check the words or phrases that apply to you.

- Strong; need to get things done NOW!
- Easily rattled by workload
- Very active, almost hyper-active
- Impatient and easily frustrated
- Everything is a high priority
- Needs to keep moving; motion-machine
- Hard to relax
- Fingers in everything; likes control
- May seem demanding

- Laid back with less need to get things done today
- Not bothered by workload pressures
- Stable and steady work pace
- Calm and even-tempered
- Only a few, clear high priorities
- Few wasted movements and actions
- Relaxed, even in tough conditions
- Delegates effectively; trusts employees
- Relaxed and low-key with the team

More check marks on the left? You are Type A. More on the right? You are Type B. Type A's are frustrated that change gets in the way of getting things done in a hurry. Type B's continue with the steady pace of their work no matter what.

How does your ability to handle stress affect your ability to change?

67 Accountability: Combat The Stress Of Change

Basic Stress Management Techniques Are Essential During Change.

Know Your A-A-B-C's

A = Awaken Your Awareness—tune into feelings; ask yourself, "What do the feelings mean?"; meditate; journal; pray; talk to a friend; seek support in a group; practice deep breathing; use progressive muscle relaxation; practice yoga.

A = Adjust Your Attitude—change your mind; explore other options for how to see things; reframe the situation; ask yourself, "How else could I see this?"; ask others the same question; deliberately look at positive things; watch a funny movie or spend time with a funny friend.

B = Become Physically Stronger—eat well; exercise; get enough sleep; stop smoking and drinking to excess; take vitamins; eat breakfast; take breaks; walk more; lift weights; do all proper physical maintenance.

C = Control What You Can, Accept the Rest—solve problems that are solvable…NOW; manage time well; set priorities and do first things first; confront issues that deserve confronting; ask for resources to help you succeed; focus on strength and successes.

What A-A-B-C can you use to manage the stress of change?

68 Accountability: Flexibility Quotient

How Flexible Are You?

Take this quick assessment to find out.

- Highly conscientious; do homework; well prepared
- Follow systems and procedures to the letter
- Love predictability and control; no surprises
- Dislike confusion and disorganization; flustered in chaos
- Plan it

- Enjoy the gray zone; like plunging into new things
- Work well with less structure; fewer procedures
- Enjoy living moment to moment; with many surprises
- Like improvisation and spontaneity; thrive on some chaos
- Wing it

More check marks on the left? You are less flexible and prefer a more black and white world. More check marks on the right? You are more flexible and roll with the punches during change.

How could you be more flexible during change?

YOUR NOTES:

69 Accountability: Flexibility Stretch

Here Are Some Ways To Become More Flexible.

Flex your Flexibility Muscles by doing these things.

- Stop insisting on "shoulds" and "should nots" and focus on "what is."
- Move beyond "right" and "wrong" to "what else is possible?"
- Instead of stating your opinion, ask a question. "Why do you think that?"
- Cultivate a problem-solving mind instead of a judgmental mind. "How can we fix this?" vs. "Here is my opinion…"
- Learn to embrace a bit of clutter and disorganization.
- Find small wins every day—your own and others'.
- In the face of failure, ask, "Where are the opportunities?"
- For every mistake and set-back ask, "What can we learn from this?"
- Expect interruptions, mistakes and setbacks instead of perfection.
- Craft plans that include interruptions, mistakes and setbacks, not perfect execution.

- Understand that changed plans, unexpected hurdles, and small failures are part of the price of progress.

What will you do this week to become more flexible?

70 Accountability: Only The Resilient Thrive

Understand Resiliency Theory And Become Stronger Through Change.

Psychologists tell us there are 5 characteristics of the resilient. During change, the resilient don't just survive. Somehow, they manage to thrive. So, for those who have not mastered thriving and are just barely surviving, here are the 5 characteristics to strengthen.

1. Build a Resilient Attitude: "I can empower myself by how I change my mind."
2. Look Long-Term: "What are the long-term gains?"
3. Practice Flexibility: "Okay, Plan A isn't working. What's Plan B?"
4. Make Order From Chaos: "What can I do? First, second, third? Every day?"
5. Recognize The Opportunities: "I see opportunity, not just danger."

Which of the 5 resilient characteristics do you do well?

Which do you want to strengthen? How?

YOUR NOTES:

71 Accountability: How Resilient Are You?

Here's A Quick Quiz To Check Your Resiliency Level.

Take a few minutes to step back and reflect on the big question and the ones that follow. Rate your Resiliency Level on a scale of 1-10 for each question.

1=What Resiliency? Not Much 10=Solid as a Rock Resiliency

- I jump in and start positive action when a change is needed.
- I actively look for ways to improve my performance.
- I build on the ideas of others for the sake of achieving a goal.
- I coach myself to a positive attitude when my spirits lag.
- I persist even when faced with obstacles to progress.
- I look for new opportunities, improve things and succeed every day.
- I keep my sights on what's important and am not overly distracted by doing only what's urgent.

What is your score for each item? Why?

What are your strengths?

What are the items you'd like to improve?

What actions will you take to improve?

72 Accountability: Your Innovation Score

Do You Dream Or Are Your Feet Firmly Planted On The Ground?

How innovative are you? Gain some insight from the following questions. Check those that apply from either column.

• Prefer doing to thinking	• Like to dream more than do
• Appreciate established methods	• Seek excitement through variety
• Practical and realistic	• Inspired by possibilities
• Better at implementing and getting things done	• Better at starting than finishing
• Sensible, grounded, and active	• Inventive, creative, and reflective
• Focus on yesterday and today	• Focus on tomorrow
• Do what's proven	• Look for ways to innovate

More checks on the left? You prefer more proven methods; change is good if the case is solid, but it had better be solid. Lots of checks on the right? You are an innovator and enjoy experimenting and improving. In short, it's easier for you to appreciate change, even if on a bit of a hunch.

How can you manage change, given your innovation score?

73 Accountability: Put On Your Innovator's Hat

Anyone Can Become More Innovative, But It Does Take Practice.

Want to become more innovative? Practice these things:

- Make of list of things that "bug" you, pick one, and fix it.
- Collect new ways of doing things from the experts; implement one or two.
- Collect new ways of doing things from co-workers and employees; implement one or two.
- Ask yourself, "How many other ways could this be done?"
- Include at least one new idea in every project.
- Remind yourself that there is seldom only "one right answer" to anything. Examine all options before deciding on a best one.
- Purposely seek opinions that are different than your own. Draw those out.
- Make a list of "sacred cow" activities in your team or organization. Challenge whether they are worth continuing.

- Find an innovation mentor and practice creative thinking skills. (see von Oech's books, *A Whack on the Side of the Head* and *A Kick in the Seat of the Pants*)

What is one practice you can use to strengthen your innovative ability?

74 Accountability: Are You A Risk-Taker?

Take This Assessment.

Some tolerate risk more than others. See where you stand. Check those that apply to you.

- Conservative and traditional
- Respect the status quo
- Play it safe
- Skeptical of new methods
- Stick to the tried and true
- Generally accommodating
- See mistakes as failure

- Visionary and "forward-looking
- Critical of the status quo
- Live on the edge
- Open to new methods
- Like to try new ways
- Enjoy challenging ideas and debating positions
- See mistakes as part of learning

Do you have more checks on the left? You are less likely to enjoy risk. Stretch by embracing risk more often. Do you have more checks on the right? You thrive on more risk-taking, less security. The risk of change is somewhat exciting to you.

How does your risk-taking profile affect your ability and willingness to change?

YOUR NOTES:

75 Accountability: Steel Your Nerves About Risk

If You're Used To Comfort And Security, Risk May Seem Foreign.

It doesn't have to be. To better tolerate risk, do more of the following:

- Engage in more conflict. Express your candid opinion. Disagree more often.
- Play the Devil's Advocate. Argue respectfully and with purpose. When views clash, ask yourself, "What is the other party trying to say?"
- Challenge rumors by asking, "How do we know this to be true?" -or- "What is the evidence?"
- Pick a small risk. Something that is uncomfortable for you but is not truly dangerous. Take the risk and notice the results. Was it really so bad?
- Do something you have longed to do but are afraid to do. Get the right support. Take a class. Get a coach. DO IT!

How will you strengthen your risk-taking ability?

YOUR NOTES:

TELL CHANGE STORIES

Stories inspire. Stories persuade. Great leaders use them and for good reason. Enjoy this section as it seeks to teach you how to do the same and provides some actual stories to use as well.

YOUR NOTES:

76 Tell Change Stories: Northern Pike

A Fish Tale For A Change

Look to the world of fish for some change wisdom.

The World Of A Northern Pike

There is a famous story in change circles of the Northern Pike who was placed in a fish tank along with some juicy-looking minnows. Unfortunately for the Pike, a glass partition separated him from the minnows.

Every time the Pike attempted to enjoy a minnow treat, he bumped up against the partition. Being a persistent creature, the Pike tried to catch minnows many times. Until finally one day, he gave up.

When the day came that the partition was removed, the Northern Pike made no attempt at all to catch and enjoy minnows. He had learned his minnow lesson awhile back and did not think to challenge it. Minnows, the Pike had accepted, were not a treat meant for him.

In what instances do you behave as the pike?

What old limitations might be left behind today?

YOUR NOTES:

77 Tell Change Stories: A View From The Window

What's Your View?

It's all too easy to think the problem belongs to the other party.

A View From The Window

Each morning the old woman woke up, made coffee and looked out the window at her neighbor's home. And each day, at that very moment, the complaints would begin.

"Why isn't their laundry cleaner? Look at how dingy it is." "Look at the dirt on that house! I wish they would clean things up. They drag down the neighborhood." "Those poor children, see how dirty their clothes are. Look at the webs in their hair!"

And each day, the old woman's husband would cringe at every complaint. They hurt his ears. Finally, one morning, the old woman sat down with her cup of coffee and began the litany of criticisms. But no sooner had she started that she stopped short.

"Why everything next door looks bright, shiny and new again! What happened?"

The husband offered, “Old Woman, last night—after you had gone to bed—I washed our window.”

In what ways might you wash your windows?

78 Tell Change Stories: A Parable

An Old Man, A Boy, And A Donkey

A parable about people pleasing.

There once was an old man, a boy and a donkey. They decided to trek to town. So off they went. The boy rode the donkey and the old man walked alongside the other two. While traveling merrily along, the three overheard some people commenting:

"Such a shame to see an old man walk while the boy rides!" They thought the comment might be right, so the boy walked and the old man rode. Not far down the path, the three overheard some people commenting:

"A grown man rides comfortably while a little boy walks? So sad!" So the old man and the boy both walked for a while, until they overheard:

"How foolish to walk when there is a perfectly good donkey to ride!" So the old man and the boy both rode the donkey, until they overheard:

"How cruel for two to ride a small donkey!" That's when the old man and the boy decided to carry the donkey, and as they crossed a bridge, lost their grip on the donkey, which plunged into the river and drowned.

The moral? If you try to please everyone, you can kiss your ass good-bye!

How can you listen with sensitivity during change without trying to please everyone?

79 Tell Change Stories: Change In A Jar?

The Jar Has It!

A company introduced a new process to improve efficiency and reduce errors. But the errors continued, despite the new and improved way to work.

The cause? The workers did not understand the real impact of the errors, which resulted in returned shipments.

The solution? A consultant placed a pickle jar in the middle of the floor. For every successful shipment, management put a quarter in the jar. For every mistake, management removed $10. The money left at the end of every month was given to the people on the floor.

The result? Within a month, errors went down and morale went up.

The moral? What gets measured gets managed. But especially if the measurement is visible, tangible, and meaningful.

If you want to direct people's attention to the change, measure what matters.

How do we measure what matters during this change?

YOUR NOTES:

80 Tell Change Stories: A Bible Parable

The Farmer And His Field

A Bible parable for work. Matthew 13:1-9, 18-23.

Some of you may remember Jesus' story of the farmer and his field. It goes something like this. A farmer plants his field and:

- Some seeds fall on the path where the birds promptly eat them.
- Some seeds fall on the rocks and without proper soil, nourishment and water, they die.
- Some seeds begin to grow amongst the weeds, are smothered, and die.
- But some seeds do land on fertile soil, take root, thrive and grow into a grand harvest.

What is the point? Well, there are many potential points, but try this one on for size. Plant lots of seeds. Expect some to grow. Plant in fertile soil for a full harvest.

Translated: To facilitate successful change, communicate many messages to many people and many times. Craft them all carefully, knowing that communicating a message one way and one time is not enough. Be careful to plant many of those messages with the organization's Opinion Leaders. Use several different messages to make sure to get through. Some of the seeds you plant will make the harvest plentiful.

How many seeds for change have been planted?

Are they on fertile ground?

81 Tell Change Stories: A Modern Parable

The Parable Of A Train In A Storm

One student's experience on a train…

A college student—traveling on a train during a blizzard—held a conversation with the engineer. He asked the engineer how he felt guiding the train through such a seemingly dangerous flurry. The engineer replied:

"Look at the engine headlight. Doesn't that light up the track for a hundred yards or more? Well, all I try to do is to cover that hundred yards of lighted track. That I can see, and for that distance I know that the way is safe. And, believe me, I have never been able to drive this old engine of mine—God bless her!—so fast as to outstrip that hundred yards of lighted track. The light of the engine is always ahead of me!"

During change—during life, for that matter—we may not know what lies ahead. But for a few yards, or possibly a few feet, the track is clear. We know the next right step. And, if we are wise, we take the step even when the whole of the journey is not certain.

What are the steps you need to take for the next hundred yards?

YOUR NOTES:

82 Tell Change Stories: A Fish Story

A Fish Who Doubted

A fish story for change.

Two fish were swimming when they saw a piece of meat dangling in the water. The younger fish darted toward it eagerly, prepared to have a meal. The older fish cried out, "Stop! You can't see it, but there is a hook inside that meat. The hook is connected by an invisible line to a pole outside the water. There is a man holding the pole. If you eat the meat, the hook will catch in your jaw and the man will pull you out of the water and eat you." The younger fish was saved...for the moment.

But later, the younger fish said to himself, "I wonder if my friend was really right. I must see this for myself." He went back to the meat. He swam far and wide around the meat, looking for a pole or a man. He could find none. So, the younger fish concluded that all was safe. He gulped the meat and was caught by the hook. Well, you know the rest. Younger fish was a fine meal.

We can look far and wide during change, trying to discover what those at the top may see, or what our more experienced colleagues may know. But if we do not see things from their angle, no amount of searching will enlighten us. If we only look at things from our own perspective, our sight, while valuable, is still a bit narrow and limited. We need the perspective of others to complement our own.

What do others think, see, or know that you do not and might learn from?

83 Tell Change Stories: Opportunities

Obstacles, Obstacles

When You Meet An Obstacle, Do You Move It?

Once upon a time, a king placed a boulder in the middle of a road. Then he hid himself and watched to see what would happen. Many travelers—wealthy merchants and solid citizens of the kingdom—walked AROUND the boulder, some loudly complaining about road conditions. But no one even tried to move the boulder.

Along came a mere peasant with a load of vegetables. The peasant stopped and, after much effort, the peasant moved the boulder. Just before resuming his trip, the peasant saw a purse lying in the road where the boulder had been. He opened it. The purse contained gold coins and a note from the king giving the one who moved the boulder permission to keep the gold. The peasant continued along the road, a rich man.

Obstacles can be opportunities in disguise. Keep that in mind as you struggle through certain aspects of change.

What are some opportunities disguised as obstacles that you have discovered during a change?

YOUR NOTES:

84 Tell Change Stories: Recognize Gifts

The Boatman Who Could Swim

Everyone has gifts. Recognize them. Value them.

Some scientists attended a convention at a resort on a lake. A mathematician, a physicist, an astronomer and a molecular biologist hired a boatman to row them around on the lake. The scientists discussed many things, from String Theory to Bubble Universes. The boatman looked curious. The physicist asked, “What do you think of these ideas?” The boatman replied, “I don’t understand any of it as I did not go to school of any kind.” “I hate to say it,” said the physicist, “but you seem to have wasted a good part of your life.”

The boatman rowed on in silence. Suddenly, a storm descended and the waves caused the boat to capsize. The boatman started swimming for shore. The scientists cried out, “Help! We can’t swim!” The boatman called back, “I hate to say it, but you seem to have wasted your whole lives.”

Everyone’s gifts are necessary for successful change. Value them.

What gifts do you bring to the change?
Others?

YOUR NOTES:

85 Tell Change Stories: A Frog Story

The Frog In The Pail

Keep on keeping on...

A frog was hopping around a farmyard and fell into a pail half-filled with fresh milk. The frog jumped hard, trying to escape the pail. He did not succeed.

Determined not to give up, the frog continued to struggle. He kicked and squirmed furiously, and in doing so, churned the milk into a big hunk of butter. He leaped from the top of the butter and right out of the pail.

Never give up. The struggle may be challenging and a strain, but it's worth it.

What are you tempted to give up?

How will you persist through the struggle?

YOUR NOTES:

86 Tell Change Stories: What Do You See?

A Man And His Ax

Sometimes we see what we want to see.

A man lost his ax. He suspected his neighbor's son of stealing it. The way the boy walked, looked and talked made him look guilty of the crime. Or so the man thought. But one day the man found his ax in the cellar. Apparently, the neighbor's son did not steal the ax after all.

It's easy to see what we've made up our minds to see. We don't believe it because we see it. We see because we believe it. During change, we must be careful to see everything, even those things that defy our beliefs.

What have you seen because you believed it to be true, even when it turned out to be false?

YOUR NOTES:

87 Tell Change Stories: Creation:

A Sioux Indian Story

Change really is what you make of it.

The Creator said, "I want to hide something from the humans until they are ready for it. It is the realization that they create their own reality." The eagle said, "I will take it to the moon." The Creator said, "No." The salmon said, "I will take it to the bottom of the ocean." "No." The buffalo said, "I will bury it in the Great Plains."

The Creator said, "No, none of these will work." Grandmother Mole, who lives in the breast of Mother Earth, and who sees with spiritual eyes, said, "Put it inside of humans." And the Creator said, "It is done."

We DO create our own reality. That is not to say that we control everything. We most certainly do not. But we DO control our reaction to everything we experience. So, the thing to change is our own—reaction, that is.

What is your reaction to change?

How could you change your reaction?

YOUR NOTES:

88 Tell Change Stories: Do Something!

Do Something!

Don't just gripe and moan.

A man looked around, saw all the pain and suffering of the world, and agonized. He yelled at God. "Look at this mess. Look at all this pain and suffering. Look at all this killing and hate. WHY DON'T YOU DO SOMETHING!" God spoke to him. "I did. I sent you."

Sometimes change looks so threatening that we want a rescuer to make it all better. Why not decide to make it better yourself? Do what you can, every day, to help yourself and others through the change.

What can you do to help people through the change?

YOUR NOTES:

89 Tell Change Stories: Not Easy To Understand

Have A Beginner's Mind

It's not so easy to understand when you're already so sure of the answers.

Once upon a time, a student went to the Zen master to learn the meaning of Zen. The Master quietly poured a cup of tea. The cup was full but the Master continued to pour anyway. Tea flowed everywhere. The student impatiently complained "Why do you keep pouring when the cup is full?"

The Master said, "I am showing you how your mind is so full of your own opinions, there is not much room for understanding Zen. Please empty your mind of your preconceived ideas and I will be able to teach you."

We have strong opinions because of our past experiences. Good. But we also need to be aware that our strong opinions can get in the way of learning new things. If we have enjoyed success yesterday based on certain understanding, who says that tomorrow's success will be based on the same?

Is your teacup overflowing? If so, how so?

YOUR NOTES:

90 Tell Change Stories: Student Story

Lessons Learned

An important lesson as told by a student.

I once had a professor who changed my life. I breezed through a test until I got to the last question. "What is the first name of the woman who cleans the school?" I was shocked. This had to be some kind of joke. I had seen the cleaning woman before. She was tall, dark haired and in her 50's, but I did not know her name.

I handed in my paper, the last question blank. A student asked if the last question would count. "Absolutely," said the professor. "You will meet many people during your lives and careers and they are all important. Remember that." I've never forgotten that lesson. I also learned the name of the cleaning lady. Dorothy.

We can be competitive. We are taught that. Who could blame us? But, sometimes our competitive training keeps us from recognizing all the important contributions of some people in some jobs. Don't let that happen to you.

Who are the "cleaning people" in your organization?

How can you recognize them during this change?

YOUR NOTES:

91 Tell Change Stories: Ask For Help

Moving Rocks

Are you willing to ask for help...when you need it?

A little boy was spending his Saturday morning playing in his sandbox. He had cars and trucks, pails, and shovels. While playing, the boy found a large rock in the middle of the sandbox. The boy managed to dislodge the rock. Struggling, he pushed the rock across the sandbox with his feet. However, the boy couldn't roll the rock over the little wall.

The determined little boy tried hard, but every time he got close to success, the rock tipped and fell back into the sandbox. Finally the boy burst into tears. Just then, a large shadow fell across the boy and the sandbox. It was the boy's father who said, "Son, why didn't you use all the strength that you had available?" The boy argued, "But I did, Daddy!" "No son," the father replied. "You didn't ask me." The father removed the rock from the sandbox.

Sometimes we need help during change. Others can do what we cannot easily accomplish. Better to ask for help than to struggle along all alone.

What are your rocks during change?

Who can you ask for help to remove them?

YOUR NOTES:

92 Tell Change Stories: Make The Decision

The Story Of 3 Frogs

Make the decision.

That's good. But action makes the decision complete.

QUESTION: Once upon a time there were three frogs on a log. One of the frogs made a decision to jump. How many were left?

ANSWER: Three. The one frog that made a decision only made a decision. He took no action!

What are the decisions and actions you must make to support the change?

YOUR NOTES:

93 Tell Change Stories: Bad Day

A Signal For Good

Sometimes when things look bad, they're not so bad after all.

A shipwreck survivor was washed up on a small, deserted island. He prayed to God to rescue him. Every day he looked for help. No help came. He built a small hut and filled it with all that he had to survive. One day, he arrived home to find his little hut in flames. Everything was lost.

The man wailed. "God, how could you do this to me!" The next day, however, he was awakened by the sound of a ship. The ship had come to rescue him. "How did you know I was here?" the man asked. "We saw your smoke signal," they replied.

What looks like disaster today can become treasure tomorrow. Don't give up hope. During times of change, many things that look bleak today can be transformed into light tomorrow.

What looks bad today that might become good tomorrow?

YOUR NOTES:

94 Tell Change Stories: Struggle

The Beauty Of Struggle

Without struggle, we may not become all that we could be.

A man found a cocoon and decided to watch until it hatched. One day a small opening appeared and the man watched the moth for several hours as it struggled to force itself through the little hole. Then the moth seemed to stop making progress. It seemed stuck. The man decided to help the moth. He snipped off the remaining bit of the cocoon. The moth then emerged easily, but it had a swollen body and small, shriveled wings.

The man thought the moth would recover with time. It did not. In fact, the little moth spent the rest of its life crawling around with a swollen body and shriveled wings. It never was able to fly. What happened? The moth needed to struggle. The struggle was Nature's way of forcing fluid from the body of the moth into its wings so that it would be ready to fly.

To go through life and work without challenge, struggle and obstacles would stunt our growth. Like the moth—without the struggle—we might be crippled and doomed to live our lives with hampered wings.

What are the struggles of change that have helped you grow?

YOUR NOTES:

95 Tell Change Stories: Teams

Fly Like The Geese

Change is best accomplished by teams.

Why do geese fly in the "V" formation? As each bird flaps its wings, it creates an uplift force for the bird immediately following. By flying in "V" formation, the whole flock adds at least 71 percent greater flying range than if each bird flew on its own. When a goose falls out of formation, it suddenly feels the drag and resistance of trying to go it alone.

When the head goose gets tired, it rotates back in the formation and another goose flies point. Geese honk from behind to encourage those up front to keep up their speed. Finally, when a goose gets sick or is wounded and falls out of formation, two other geese follow the goose down to lend help and protection. They stay with the fallen goose until it is able to fly or until it dies.

For change to be effective, we need teamwork like that of the geese. So:

- Fly in formation, aligned with the rest of the team.
- Understand that you'll fly faster and farther with the team than alone.
- Be prepared to rotate positions when needed.

- Encourage everyone on the team, especially when you're coming from behind.
- Protect and help those who are injured or wounded during the change.

How can you be a better team player during change?

96 Tell Change Stories: Thirsty For Change

How To Sell

Understand what really moves people during change.

A young salesperson lamented losing an important sale. She had done what she believed was a nice job of explaining the features and benefits of the product she was selling. She discussed it with the Sales Manager and commented. "I guess it just proves you can lead a horse to water, but you can't make him drink." The Sales Manager replied, "Your job is not to make him drink. It's to make him thirsty."

What the sales person had failed to do was to discover the prospect's needs. She did not do a good job of discovering what the buyer has now versus what the buyer desires for the future. That "point of pain" might have provoked the buyer to be more open to the pitch of product features and benefits.

To sell change, take the Sales Manager's advice. Make people thirsty for the change. Remind them about the pain of the current way of doing things. Why is the current way not working? Show them how a new way will change their lives. For the better. What will the change vision offer?

How can you make people thirsty for change?

YOUR NOTES:

97 Tell Change Stories: Impossible

The Fish And The Turtle

Just because you have not seen it or experienced it, doesn't mean it's not possible.

Once upon a time there was a fish that had lived all of its life in the water. One day, the fish—while swimming around in the water—met his friend, the turtle. "Good day, Mr. Turtle!" said the fish. "Where have you been?" "I have just been for a trip on dry land." the turtle replied. The fish exclaimed, "Dry land? There is no dry land. I have never seen it." "Well," said the turtle good-naturedly, "there certainly is dry land."

The fish insisted, "Okay then, tell me about dry land! Is it all wet?" "No, it is not wet," said the turtle. "Is it cool?" asked the fish. "No, it is not cool," replied the turtle. "Is it clear so you can see through it?" "No, it is not clear." "Could I swim in dry land?" "You could not swim in it." "Are there waves on dry land?" "No." said the turtle one more time.

The fish exclaimed again, even more fervently, "Then there can be no such thing as dry land. Dry land is nothing." The turtle said, "If you insist that dry land is nothing, then to you it is nothing." And with that, the turtle went off for another excursion on dry land that is nothing.

They thought the world was round. They discovered otherwise. Over time. They did not believe we could travel to the Moon. Again, they learned differently. Some thought the Internet was a crazy idea. We all know what happened with that. It is tempting to believe in only what we've personally seen, heard, or experienced. How limiting! Some of the world's greatest advances once seemed impossible but turned out otherwise.

What seems impossible about this change?

How can you change your mind?

98 Tell Change Stories: Encouragement

The Parable Of Two Frogs

A tale about encouragement.

Two frogs fell into a deep pit in the woods. All of the other frogs gathered around to see if they could help. As they surveyed the situation, the crowd of frogs became more and more dismayed and concluded that the situation was hopeless. Unwilling to accept this terrible fate, the two frogs began to jump with all of their might.

The frogs looking on warned the two frogs in the pit to save their strength. There was nothing to be done. The two should consider themselves as good as dead. If they had not been so careless, they would have not gotten into this terrible situation in the first place.

The two frogs continued jumping as hard as they could, and soon became very tired. Finally, one of the frogs—swayed by the frogs looking on—gave up. He gave up and died. The other frog continued to jump with full energy, despite pain and fatigue. The frogs above continued to warn and criticize the frog's efforts. The weary frog jumped harder and harder.

Then, in one remarkable leap, the frog made it! He was out of the pit! Amazed, the other frogs celebrated and asked, "Why did you continue jumping when we told you it was impossible?"

The frog responded. “I am deaf and could not hear you. I can only read lips. I thought your gestures were to cheer me on.”

And so the frog could not hear his onlookers’ discouraging views. He thought they believed in him and wanted him to succeed. Ah, the power of encouragement!

How do you most often express yourself?

With encouragement or discouragement?

99 Tell Change Stories: A King's Tale

Peace

What does peace look like?

There once was a King who offered a prize for the best painting of peace. The King liked two of the contest entries best. One picture was of a calm lake surrounded by mountains, trees, and lovely blue sky. It was a beautiful scene. The second picture was quite the opposite. The skies looked stormy, the mountains were barren, and a torrential waterfall rushed furiously. But behind the waterfall a tiny bush was growing in a crack in the rock. And in the bush was a nest. There, in the midst of the rush of angry water, sat the mother bird on her nest...perfect peace.

Both pictures were quite brilliant but the second picture won the prize. The King recognized that real peace endures through the most chaotic, challenging circumstances.

Change brings some inevitable chaos and challenge. Learn to have peace during the storm of such chaos and challenge. That is true maturity.

How can you remain calm during the storm of change?

YOUR NOTES:

Letting Go Of The Past

Learn from the failures of yesterday. Don't just dwell on them.

Two monks were on their way home to the monastery. A beautiful young woman in a bridal gown called out for help. She needed to cross the stream and did not want to get wet. The monks—due to their vows—were prohibited from touching women. One monk, however, broke the rules and carried the woman across the stream. The second monk was very angry. "How could you do that?" he scolded. "You know we are forbidden even to touch a woman, much less pick one up and carry her around!"

The guilty monk listened in silence to the angry monk. He listened all the way home. He turned his attention to the warm sunshine and singing birds. After returning to the monastery, the angry monk kept complaining. "How could you carry that woman? You broke the rules. How could you? Someone else could have helped her across the stream. You were a bad monk!" "What woman?" the guilty monk asked. "You don't remember? The bride at the stream?" his colleague insisted. "Oh, her," laughed the guilty monk. "I only carried her across the stream. You carried her all the way back to the monastery."

We can learn from the past. We can learn from past successes and past failings. We can learn from past ups and downs. We can learn from past celebrations and tragedies. We can learn. And move on. Or we can dwell in the past. Let's resist living in the past instead of living for today. Better to let go of yesterday and live for today.

How is the change effort hindered by people's views of the past?

QUOTE THE WISE

Quotes from those who have achieved great things, are the finest thought leaders, or just have a way with words, inspire us to examine and improve our lives and work, presumably and hopefully for the better. This section of cards contains some quotes to lift you and others up during the process of change.

YOUR NOTES:

101 Eleanor Roosevelt

You gain strength, courage, and confidence by every experience in which you really stop to look fear in the face. You must do the thing you think you cannot do.

~ Eleanor Roosevelt

People know they spare themselves some anxiety by simply avoiding what is feared. Playing it safe is comfortable. Easy.

However, most people don't consider the cost of such avoidance. They forget that the price of playing it safe is: **Loss Of Confidence.**

Tackling what you fear may not result in instant success or monumental achievement. But, it's a sure bet that tackling what is feared will breed confidence. Add to that, an even more resounding outcome of facing one's fears is the most deliciously refreshing and surprising realization that you are so much more capable than you ever knew.

What fears have you faced?

How did you do it?

What was the impact on courage and confidence?

102 John Wayne

Courage is being scared to death, but saddling up anyway.

~ John Wayne

The wise have always known it. Courage is not the absence of fear. Courage is the will to act despite fear. How to do this? You can:

Coach yourself through the fear, just as you would coach your child, a friend, or a colleague through a fear. Talk to yourself as you would talk to them.

Get fully prepared for the challenge. Study. Practice. Get a teacher or coach. Take a class.

Weigh the real risks and rewards. Make a list of pros and cons. Examine the list. Focus on the rewards. Head off some risks with intelligent preparation.

Seek counsel. Talk through your fears and concerns. Get input and feedback. What's realistic? What's not? What can be done to raise the odds of success?

What can you do during this change to act, despite fear?

103 Peter Drucker

Unless commitment is made, there are only promises and hopes...but no plans.

~ Peter Drucker

Commitment. One dictionary definition is: A pledge to do.

Another definition: The state of being bound emotionally or intellectually to a course of action or to another person or persons.

Pledge to do. Being bound to a course of action. Commitment sounds simple. But it's not. Organizations often lack commitment exactly when they need it most. How do we know? Things are promised but don't happen. Serious issues get left alone when they would better be confronted. People speed through the obligations attached to change in order to get back to "real work."

Going through the motions does not ensure commitment. Nor does avoiding respectful conflict. Or failing to follow-through. How to deepen commitment when it's lacking? First, recognize that lack of commitment is the problem. Then, identify and resolve the causes.

Where there are lingering reservations about a course of action, discuss those. Where there is fear of failure, get better prepared for success and help others do the same. If apathy is the problem, point to the vision for the change and link it to people's day-to-day work. If past change failures have left a sting, discuss those wounds openly and point out how this change will indeed be different.

Where are signs of wavering commitment in you, the team, or the organization?

What can you do about it?

104 Pericles

What you leave behind is not what is engraved in stone mountains, but what is woven into the lives of others.

~ Pericles

In business, we look for tangible, practical results of change. Increased revenues. Breakthrough innovations. Lowered costs. Improved efficiency. Better profits. This is fine stuff. But, there are other payoffs from working through change.

Communities benefit from increased economic prosperity. Lives are improved—even saved—through innovation. The workforce acquires new skills and growing courage from facing challenges. The list goes on.

Changes like these do not show up on the financial statements, but they exist nonetheless. Remember that. Remember that, especially during those times when change is the most stressful and the evidence may not appear as writings in stone mountains.

What are the tangible and intangible outcomes of the current organizational change?

YOUR NOTES:

105 Abraham Lincoln

The dogmas of the quiet past are inadequate to the stormy present. The occasion is piled high with difficulty, and we must rise with the occasion. As our case is new, so we must think anew and act anew.

~ Abraham Lincoln

We wear our old thinking just as we wear our favorite pair of old, but exquisitely comfortable shoes. The shoes may look a bit worn and may be far out of fashion, but they still suit us. They are COMFORTABLE. Cushy, broken-in, well-worn, fitted-just-to-us, comfortable. A pair of new shoes might entice as they are shiny and in vogue. But we know that new shoes require some real wearing, oiling, and breaking in before they'll seem just right.

So it can be with new thinking. We count on our old thoughts. They are certain, or so we rely on them. The old thoughts got us to where we are today—for better or worse, although we tend to see the better side of the equation. Why give them up? New thoughts require the same real wearing, oiling, and breaking in that new shoes do before they seem quite right. The trick is to get people to wear new thinking as they might new shoes. They must wear them long enough to properly break them in.

What are some of yesterday's assumptions and beliefs that led to success?

What are new assumptions or beliefs that will lead to success tomorrow?

106 Epictetus

It is impossible for a man to learn what he thinks he already knows.

~ Epictetus

In many classes, there are those who clearly express the attitude, *"I Know This Already And Don't Need To Know More*."

People of those types are not only found in classes. They walk the halls and aisles of our workplaces every day. Too bad for the workplace!

The truth is, even the most learned among us can…**Learn More!**

In fact, ironically, those who know that they don't know, are likely to be the most learned. Really. They are open. Receptive. Ready. Beware the disease of believing that you "know it all."

What are you tempted to believe you know so well that there is nothing left to learn?

107 George Bernard Shaw

Some men see things as they are and say "Why?" I dream of things that never were and say "Why not?"

~ George Bernard Shaw

It's easier to moan about the current circumstances than to invent solutions and strategies for a better life, business, or world. It's even harder to put those solutions and strategies into action.

Listen to yourself for the next week or the next month. What do you say? Are yours the words of a complainer or an inventor? Vow to adopt more Inventor Speak in the coming year.

What do you complain about but fail to resolve?

What are some solutions and strategies you could implement?

108 Henri Bergson

To exist is to change, to change is to mature, to mature is to go on creating oneself endlessly.

~ Henri Bergson

We fool ourselves into thinking that we have "arrived." We believe that after a certain number of years, we have learned our craft. We believe that after earning a certain amount of money, we have found security. We believe that after spending a certain number of years with a company, we can simply coast. Well, maybe.

In reality, the world is constantly changing and those of us who wish to mature, we will go on changing also. On purpose. By design. We will resist believing that we have ever "arrived."

Not that we should not respect our accomplishments or look back with gratitude, but after celebrating those successes, let's set some goals to start the change again. Let's take charge of the change, not be simply subject to it or an observer of it.

What is a change that will take you to the next level of maturity?

109 James Yorke

The most successful people are those who are good at Plan B.

~ James Yorke

Set the goal. Make plans to achieve the goal. Commit to the plans. Get on with the action. But be ready to go to Plan B. Life does not guarantee that Plan A will always work out. Plan B may be the way to success.

Just because you committed to Plan A does not mean that Plan B is not viable and deserving of an equal share of commitment. Be smart. Be flexible. Do not see a departure from Plan A as a failure. Do see the shift to Plan B as intelligent.

When did you, the team, or the organization successfully shift to Plan B?

110 Nelson Mandela

Education is the most powerful weapon which you can use to change the world.

~ Nelson Mandela

An education is the most powerful weapon you can use to change businesses and organizations. Educate people so they are ready for the change. Offer training about the technical aspects of the change. Hold classes about how to respond to change itself.

Create forums for questions and answers—a more informal educational process. Put updated change information for all to learn from on the web. Make a Change Question of the Week bulletin board for impromptu questions that deserve an answer—publicly, to many, in real time.

Use a variety of avenues to broadcast information in order to make everyone smarter. In the end, if the change itself does not achieve what was envisioned in the first place, you'll still have a whole lot of smarter people in your team or organization. Isn't that alone worth it?

What are all the ways we can make the people involved in this organization change smart?

111 Thomas A. Kempis

Be not angry that you cannot make others as you wish them to be, since you cannot make yourself as you wish to be.

~ Thomas A. Kempis

Organizations can easily become filled with angry people. But angry at what? Often, it's "THEM." Management who breaks promises (or so it looks). People who just complain and moan (or so it sounds). That team "over there" who just doesn't get the job done (or so it seems).

What is this anyway? "Make others as you wish them to be?" What about inspiration? What about influence? What about starting with yourself? Here are the questions.

- What is the depth of your conviction about the change?
- How fully have you taken responsibility for the change?
- How well do you "walk the talk" of change?
- How hard have you worked to make the change happen?

Forget about "them" (or at least focus on them less). Focus on you. That's the source of your true power.

What can you do to take responsibility for the change?

112 John Welwood

The most powerful agent of growth and transformation is something much more basic than any technique: a change of heart.

~ John Welwood

Change can be a head trip. More often, change is a heart trip. Sure, it's a good idea to outline and understand the Business Case for change. There is value in knowing the facts and figures that justify why the change makes sense. How will it affect the customer base? The workforce morale? The revenues? The expenses? The bottom line? Yes, the business case for change is important.

But experts tell us—and common sense does too—that the heart case for change moves people more than the head case does. How does a change of heart happen?

- Stories
- Experiences
- Testimonials
- Quotes
- Examples

In the end, what moves people is an understanding about how change affects…you guessed it: people. Real world. Real ways. Every day. So, most importantly of all, we must understand that in addition to the business case for change, the emotional case for change is what inspires people to act. The logical case for change may guide forward but the emotional case for change powers forward.

How will the change in your team or organization affect people "up close and personal?"

113 Caroline Schoeder

Some people change when they see the light, others when they feel the heat.

~ Caroline Schoeder

Understand what motivates people. So you can motivate yourself. And then, so you can motivate others. Sufficient pain can drive almost any of us to choose a different path. So can the bright light of amazing possibilities.

So, expose yourself. To the pain. Don't hide from it. To the light. Seek it out. And move on to your own personal greatness.

What is the pain that can drive you to change?

What is the light that will attract you to change?

114 Thomas Fuller

Nothing is easy to the unwilling.

~ Thomas Fuller

When faced with the zip line—1200 feet long, 150 feet in the air—I stiffened. Three false starts and I finally let go!

Exhilaration! Down the zip line in just a few exciting and enriching seconds. Why the three false starts? Resistance. Plain and simple. Fear and the resistance that is its companion.

In the end, the false starts turned out to be far more taxing than the actual ride down the zip line. In one simple and straightforward zip-line exercise, I understood that resistance is just wasted energy. Basically wearing one out with little to show for it after all is said and done. In my case, the exhilaration, the excitement, the enrichment could have come much sooner. And easier. The resistance, while understandable, was purely wasted time and energy. Is it the same in your life?

What are you resisting that—if you just took the desired, but feared action—might exhilarate, enrich, and excite you?

115 Ralph Waldo Emerson

The reward of a thing well done is to have done it.

~ Ralph Waldo Emerson

Certainly organizations are smart to reward those who support the current change. And we all become accustomed to those obvious, external rewards. But there are rewards we often forget about that are so much more powerful than money or certificates or prizes. One of the most powerful of those rewards is the

Sense Of Satisfaction Of A Job Well Done.

Don't kid yourself. Even the most cynical among us feel some pride from work done well. The feeling of exhilaration from knowing you made the right contribution in the right way cannot be replicated by any external reward. So, go for the rush. Do the work well. You'll be glad you did.

Describe a time you have enjoyed the exhilaration of a job well done?

116 Chinese Proverb

You cannot satisfy hunger by drawing a cake.

~ Chinese Proverb

Okay, you can inspire people to move forward with the change vision. The vision works wonders for sparking people's energy. But, after a time, the change vision wears thin…without some results to back it up. So, show them results.

Tell change success stories. Offer change results. Quench their thirst for change water.

How can we not only draw the cake, but give them cake?

117 Chinese Proverb

Genuine gold fears no fire.

~ Chinese Proverb

Feedback is not so easy to hear. No matter. It's important to hear anyway. If you're the REAL DEAL—a leader with substance—you'll do the drill. You're real gold, so you don't fear the fire.

Ask about how the change is going. Ask about obstacles. Ask about success stories. Ask about what else needs to get done… that's not getting done. Listen carefully. Ask more questions, seeking clarification. Resist the urge to defend. Here are some key questions. Ask them…often.

What's working with this change?

What's not working?

What are the biggest obstacles?

What are some success stories?

118 Ken Keyes

Did all your uptightness, fear, anger, jealousy, worry, resentment, grief, irritation, and heartbreak solve your problems?

~ Ken Keyes

Okay, change stirs up emotions. For many, the immediate response to change of any kind is "fight or flight." In short, survival. On an emotional level the change is treated as if it's "life or death." Even as time goes on and the change unfolds, emotions can be strong. The question is, will we put ourselves in charge of our emotions or will our emotions be in charge of us?

So, feel the feelings. Then, step back from them. Ask yourself:

What does this emotion mean?

What is it teaching me?

Then, move on to the next right step—fear, worry, resentment—or not. Don't "act out" your feelings. Examine them and act in accordance to what is the right next step, no matter how you feel. Right action, not just reaction.

119 Maya Angelou

If you don't like something, change it. If you can't change it, change your attitude. Don't complain.

~ Maya Angelou

Change or complain. Which do you choose? In a strange but real way, complaining simply perpetuates the way things are. Complaining maintains the status quo and therefore, does not change a thing for the better or even for the worse. So, if you claim to earnestly want things to be different, complaining clearly won't get you there. Why bother with complaining at all then? The answer is simple.

Complaining Is Easier. Less Risky. The complainer's path is the Path Of Least Effort. But, least effort = least reward.

Oh, the lure of the temporary release of anxiety immediately following complaining is strong. Don't take the bait. The way to real, lasting, powerful, positive change is far more satisfying. Take on the real change; not just the fake variety, aka, the complainers choice.

What is a complaint you could transform into a change for the better?

YOUR NOTES:

120 W. H. Murray

Concerning all acts of initiative and creation, there is one elementary truth—that the moment one definitely commits oneself, then Providence moves too.

~ W. H. Murray

Sometimes we complain that a change is not coming our way, or is not coming our way fast enough. Maybe WE are in the way. If we go after a challenge half-heartedly, with unresolved reservations, or a vision not yet fully formed, we are likely to stumble over our own feet. Knowingly or unknowingly.

As mysterious as it seems, when our commitment to any change—large or small—is full and complete, things that stand in the way of the change seem to fall away, almost as if by magic. And all the resources, ideas, and support to make the change a reality seem to fall into place, relatively easily and somewhat mysteriously.

Make the decision!

Make it deeply. With your complete heart. Oh, you may still have some doubts from time to time. Or some change challenges. But no matter. The commitment will carry you through.

How do you know when you are fully committed to a change?

How can you build commitment to this change?

121 Anne Morrow Lindbergh

Only in growth, reform, and change, paradoxically enough, is true security to be found.

~ Anne Morrow Lindbergh

So you think your security comes from being firmly and solidly entrenched where you are today? With the same projects? The same skills? The same people, places, and things? The familiar? Well, it may seem so, but that's an illusion.

True security comes to those who spread their wings. Find new challenges. Acquire new skills. Develop new abilities. True security comes to those who seek the UNKNOWN.

What are your opportunities to seek the unknown?

YOUR NOTES:

122 Dr. Robert Anthony

When you blame others, you give up your power to change.

~ Dr. Robert Anthony

Have you ever heard the saying?

When You Point With One Finger, Two Are Pointing Back At You.

It's true. Sure, others may do things we don't like. In terms of change, projects get proposed and started without our full input. Or with little discussion. Or not enough information. The natural urge is to resist. And, feeling powerless and fearful, we complain. It's seems as if complaining is the magic salve. Quick. Easy. Calming (in a twisted sort of way). Handily within reach.

Trouble is, complaining offers only a fleeting sense of relief. It just doesn't last. So, we need to complain more. And more. Complaining almost becomes a "fix." We can become complaint junkies.

To get a more lasting sense of satisfaction and boost your personal power, take more positive action. Ask questions. Offer ideas backed by facts and information. Share your feelings, but do so responsibly. Discover what the change really means in your part of the business. Sign up to learn the skills necessary to carry you through the change.

When have you pointed the finger and what was the result?

What would be a better approach to resolving anxiety about change?

123 Henry Kissinger

The task of the leader is to get people from where they are to where they have not been.

~ Henry Kissinger

To get others from where they are to where they have not been before—to where they could be—it helps to have done so yourself. In that way, you'll understand what it feels like to delve into the unknown, carrying the weight of a potful of fears. You will develop empathy for those who shrink from change, while at the same time carrying the conviction that change is just what the doctor ordered, fears accounted for, yet damned.

While moving through any change, then, develop a keen awareness of your own experiences, knowing that what you learn today might enable you to help others later. The best coaches harken back to their own play or performance—and their thoughts, feelings, and actions throughout—to know how to instruct others to achieve the same and even more.

What have been your experiences with change?

How can those experiences help you help others through change?

124 S. L. A. Marshall

The art of leading, in operations large or small, is the art of dealing with humanity, of working diligently on behalf of men (and women), of being sympathetic with them, but equally, of insisting that they make a square facing toward their own problems.

~ S. L. A. Marshall

The best leaders have empathy AND strength. They can—and do—put themselves in their followers' shoes.

They also guide followers to face performance problems. Or stretch to greater performance heights. In other words, great leaders empathize AND challenge. These are not mutually exclusive practices. BOTH are critical to the success of people.

How do you empathize with others at work, at home, or anywhere?

How do you challenge others to be their best?

125 Thomas Jefferson

In matters of style, swim with the current. In matters of principle, stand like a rock.

~ Thomas Jefferson

Change for change's sake. Not so good. Change based on rock-solid principles. Fabulous!

Even the study of the best companies shows the wisdom of this. The companies sited in *Built to Last* and *Good to Great*, sought and seek constant improvement but stick with some strong core principles. Some call the principles core values.

What's good for a high-performing company can also be practiced by the high-performing individual or team.

Solid At The Core. Flexible Around The Core.

When are you rock solid?

When do you swim with the current?

YOUR NOTES:

CHANGE FOR TEAM SUCCESS

Actually, this section might be better titled, "Team Success Creates Change Success." Getting all aligned on a common vision is key to change success. And, what better definition of team is there but "a group of individuals aligned on a common goal?" So, one can easily see that team success is integral to change success.

With those connections in mind, this section describes expert views and researched results revealing what it takes to succeed with a team, whether you are a team member or a team leader. Follow along for tools, techniques, and bits of wisdom to help build top-performing teams and ultimate change success.

YOUR NOTES:

126 Team Change: A Definition

What Is It?

Here's a broad definition of "team":

A group of 2 or more people,

who must coordinate their efforts,

to achieve a common goal.

Or, consider these synonyms:

Join, unite, work together, couple, cooperate, link up, get together, yoke, band together, collaborate, join forces, bunch, posse, company, crew, troupe, squad

Most of us know the story of a great team. Some of us have been on one.

Tell the story of a high performing team. The team can be famous or ordinary. You might have been part of the team. Or not. The team might have been a sports team, community team, business team or any other kind of team. Whatever the case, just make sure to tell the story of the team in as much detail as you can.

What does a high performing team look like?

What does a high performing team do differently from a mere group?

127 Team Change: Success Skills 1

Common Goal

Common goal. What does that mean? Let's look at each word.

Common = Bound together by a goal each team member believes in; otherwise known as: working toward the same aim; rallying around the cause; all for one and one for all!

Goal = The end result of the performance; otherwise known as: vision; mission; aim; objective; purpose; destination; finish line; desired state; results.

Great teamwork begins with the common goal. The common goal gets the attention. It sets the direction. It's the Holy Grail. Call it the ultimate destination of the whole team trip. There are 2 main things to keep in mind about Common Goal.

It must be CLEAR. You've got to see it to believe in it.

It must be MOTIVATING. The goal should be a challenge or meaningful to rally people around it.

What is the common goal for your team?

Is it clear?

Is it motivating?

YOUR NOTES:

128 Team Change: Success Skills 1

Common Goal

When a Team is not Working...

Teamwork experts, Carl Larson and Frank LaFasto—*Teamwork: What must go right/What can go wrong*—tell us that when a team is NOT effective, the most likely culprit is the common goal. Here's what can go wrong:

- **Lack of focus:** the goal is too vague, too broad, or too abstract.
- **Political interests:** other interests and agendas—inside or outside the team—compete with the goal.
- **Lack of urgency:** the team isn't excited about the goal.
- **Lack of meaning:** the team doesn't believe in the worth of their effort.
- **Unclear priorities:** too many competing goals or activities
- **Personal interests:** personal interests trump the team's goal.

So, when your team is not working, review the list and ask:

What is distracting the team from the common goal?

YOUR NOTES:

129 Team Change: Success Skills 1

Common Goal

How do you know when a goal is clear?

A clear goal is easy to visualize; you can see in the mind's eye. It should be stated precisely, not too broadly. Concrete words work better than abstract ones. The tests of clarity are:

- Everyone on the team easily describes the goal in the same way.
- The team—and everyone else, for that matter—will absolutely know when the goal is achieved.

THIS: To place a man on the moon by the end of the 1960's

NOT THIS: To become leaders in space exploration

How does each team member describe your team's goal?

How will the team know when the goal is achieved?

YOUR NOTES:

130 Team Change: Success Skills 1

Common Goals

Challenge and Meaning

Each February, two spectacular football teams square off to compete in an amazing American event called Super Bowl. Each player on each team knows the goal. Each player on each team is motivated by the challenge.

Why is it all such a thrill? There are plenty of possible answers to that question but one reason stands out from the rest. The Super Bowl offers an opportunity to test the limits. What limits? The limits of talent. Skill. Preparation. Endurance. Heart.

In a similar way, when a natural disaster occurs, it's typical to see people pull together for the rescue and rebuild. They rally to the cause. They bring lots of energy and ingenuity to really tough circumstances.

Why? It matters. It matters a lot. People's lives are at stake in a whole host of ways. Sometimes life and death are imminent. Sometimes property and the destruction of safety, security and living conditions are threatened. Sometimes an entire way of life may be at stake.

To recap. Challenge is motivating. To rise to a significant challenge, to test one's limits, to surpass prior performance are challenges that arouse our finest energies in a positive direction.

Similarly, meaning is motivating. To make a difference, to add value, to change lives, to save lives are all ways to add meaning and therefore raise spirits to do better, be better, and achieve more.

CHALLENGE & MEANING: Two ways to make the common goal motivating.

What makes your team's goal motivating?

131 Team Change: Success Skills 1

Common Goals

Distractions Can Cause a Team to Fail.

- "We seem to be more interested in deciding who's in charge than in solving the problem."
- "This is the right thing to do, but what if management disagrees?"
- "He's not going to stick his neck out if the boss doesn't like it."
- "She's with us right now, but if push came to shove, I think she'd abandon ship."
- "Everyone seems to be waiting for the whole thing to fail and things to go back to business as usual."

Distractions can cause a team to underperform at the very least and fail at the very most. When the common goal takes a back seat to any of the above—when the goal moves to second place, or perhaps even third or fourth place—a team will be in trouble. If your team stalls, it's wise to ask:

What is the team putting ahead of its common goal?

YOUR NOTES:

132 Team Change: Success Skills 2

Cohesive Strategy And Structure

The common goal—the inspiring vision is—a great start. But still, by itself, the inspiring vision is not enough. So, after the vision is set, what's next?

The team needs strategy and structure. And the strategy and structure must be aligned with the common goal, creating a tight cohesion that ensures success. According to Larson and LaFasto, teams can be divided into 3 categories. How to assign a category? Look at the broad purpose. Here are the 3 types of teams, labeled that way as each type is obviously different, based on the nature of purpose.

Tactical: A team that needs to execute things. On a dime. Little discussion. Lots of efficiency. Think fire fighters. Or many military teams.

Problem-solving: A team that must solve a current problem. Fix things. Make things better. Lots of discussion. Lots of sharing. Think rooms with people brainstorming.

Creative: A team that must invent something. From scratch. Break-through. Not the norm. Not business as usual. Instead, business-as-unusual. Think product development like McDonald's Chicken McNuggets team.

Which type of team are you? Strategy and structure must follow type—deliberately and by design. Not just be accident. So, decide your primary purpose.

**Note: Many teams believe they possess more than one purpose out of the three. We still say, one purpose ought to be primary above all.

What type of team are you?

How do you know?

133 Team Change: Success Skills 2

Cohesive Strategy And Structure

So, you think you're a problem-solving team. Or a tactical team. Or a creative team. Now what? How does the overall type of team affect strategy and structure anyway? Here's how.

Tactical Teams: Must be impeccably organized. No room for error. Everyone knows the job expectations. Everyone knows the procedures and can execute rapidly and impeccably. Lots of practice and preparation required. Such teams have many standardized procedures, to which everyone must adhere.

Problem-Solving Teams: Must share openly. Everything. No holds barred. Every relevant piece of information must be disclosed. Every idea must be explored. And, every reservation as well. Health conflict must trump political correctness. The team fervently believes that the whole is better than the sum of its parts.

Creative Teams: Creativity is the product of breaking traditional boundaries. So, creative teams must be free of typical organizational requirements. They must be free of customary organizational politics. They must also be free of those who embrace "that's the way we've always done it" above invention and progress. Creativity requires agility, breaking norms, thinking WAY outside the box.

How does your team's strategy and structure fit with its purpose?

134 Team Change: Success Skills 2

Cohesive Strategy And Structure

What are the forms of strategy and structure your team will need to perfect?

1. **Roles and responsibilities:** Who does what, when and how? What is each team member accountable for?

2. **Communication:** How easy is it to get vital information? How accurate is the information? Do team members have the chance to communicate informally? How are issues and decisions documented?

3. **Performance management:** How is team member performance evaluated? How do team members receive feedback? How are team members held accountable?

4. **Decision-making:** How are decisions made? How are the facts that support decision-making gathered and used? What types of decision-making processes are typically employed?

So, those are four areas related to strategy and structure to analyze and address. Understand them. Discuss them. Do each one well if you want a successful team.

How effective is your team at strategy and structure?

How do you know?

135 Team Change: Success Skills 2

Cohesive Strategy And Structure

A requirement for top teams?

Everyone Must Be Accountable All The Time!

First off, each team member must know his or her job and must do that job well. Second, each team member ought to understand others' jobs in case there's a need to pitch in. Third, each team member must understand the team's plays and his or her part in those plays. Fourth, when the team stumbles or fumbles, everyone must help with recovery. Failures must be examined fully and candidly. No finger pointing. No blame. Just the understanding that:

A Failure For One Is A Failure For The Whole Team.

How accountable are your team members and therefore, your team?

YOUR NOTES:

136 Team Change: Success Skills 2

Cohesive Strategy And Structure

How important is trust to teams?

Patrick Lencioni, author of *The Five Dysfunctions of a Team*, says that trust is the foundation of strong teamwork. With trust, teams produce excellence. Without trust, teams falter.

Why is trust so important? To solve problems, and solve them effectively, teams need good information. Accurate information. Views from a variety of perspectives. Candor, not politics. The better the information used for problem-solving, the better the solutions can be.

Since the quality and reliability of information is so important to effective problem solving, the converse is also true. Use faulty information? Get faulty solutions.

How to develop trust? Communicate openly. Insist on candor. Get the views of all team members. When you sense that someone is holding back, encourage disclosure. Ask for views contrary to your own, don't assume they'll just happen. Probe for dissenting opinions. Encourage healthy dissent. Embrace that dissent when it happens. Be suspicious when it doesn't (someone is likely lying or holding back). Provide absolutely accurate and up-to-date information to all team members all the time.

On a scale of 1-10 (1=lowest, 10=highest)

What is the level of trust in your team?

How can you improve the score?

137 Team Change: Success Skills 2

Cohesive Strategy And Structure

Top Teams Spend Social Time Together.

It's a fact. Top teams also communicate informally as well as formally. Meetings are important, but so is water cooler talk.

As it turns out, what gets shared at the company picnic or in the company cafeteria or in even more relaxed settings—retreats, dinners, getaways—may be more candid and creative than communication in more formal settings.

Take away the formal business attire, the boardroom setting, the typical agenda, and the titles. Replace those with some informality and relaxation—short sleeves, tennis shoes, refreshments, couches, and a nature view. This adds up to: equality, trust, openness, ease, disclosure, honesty, openness, and bonding.

How does your team socialize?

How do you get away from the formal, the routine, and business as usual?

YOUR NOTES:

138 Team Change: Success Skills 2

Cohesive Strategy And Structure

Create a Team Memory.

High performing teams keep accurate and thorough records—a team memory. They document important issues and decisions using a proven, consistent method. Meeting minutes. Web portals. Centralized files.

Here's what you'll need: A well-organized documentation system that is easily accessed by each team member.

Where and how are team decisions documented?

How easily can each team member access the information?

YOUR NOTES:

139 Team Change: Success Skills 2

Cohesive Strategy And Structure

Top Performing Players.

The old saying goes:

A Chain Is Only As Good As It's Weakest Link.

A top performing team needs top performing players. Every player must do his or her best. High performance like that cannot be left to chance. It happens by design. It requires coaching. And that means:

Evaluation And Feedback

How do you evaluate your team members?

What kinds of performance feedback do team members receive?

How do they receive it?

YOUR NOTES:

140 Team Change: Success Skills 2

Cohesive Strategy And Structure

Quality Decision-Making.

Great teams:

1. Decide how to decide.
2. Make decisions based on facts.

First things first. For every decision, decide how to decide. The choices are:

- Executive decision?
- Consult with many, then decision by one?
- Voting?
- Consensus?

Once the team has decided how to decide, gather the facts needed to make the decision. This may not be exactly "nothing but the facts," but as many facts as possible, based on the soundest data gathering methods possible.

The quality of your decision-making depends upon it.

How does your team decide how to decide?

On a scale of 1-10 (1=low, 10=high)

Are your team's decisions based on sound facts and information?

141 Team Change: Success Skills 3

Competent Team Members

To build a strong team, pick the right people. Sounds sensible, right? Research reveals that what sounds sensible IS sensible. And, great companies put it in to practice. Take Google, for example.

Every year, about 3,000,000 people apply to work at Google.

Only 7,000—just 0.2%—get the chance.

Laszlo Bock, Google's hiring chief insists that Google's hiring practice is critical to Google's success. According to Bock,

"The number one thing [we learned] was to have an incredibly high bar for talent and never compromise. [Google's founders] Larry and Sergei figured out early on that there's a reversion to the mean when you don't. Let's say you hire a great person, they'll hire someone almost as good, who hires someone almost as good, and over time you're just hiring average people."

Picking the right people is fundamental to team success. How to do it?

Start by knowing what NOT to do. Don't pick people for the wrong reasons. Don't pick Sally because Sam (the big boss) likes her. Don't include Jim just because he likes the team topic. Don't select Ann because she needs to "buy in" to the team's solution. Such reasons—while nice, sensitive, or convenient—don't lead to team success.

So how should one pick the right team players? Follow this rule:

Pick The People With The Competence To Achieve The Common Goal.

Review the common goal and ask:

What are the knowledge, skills and abilities critical for success?

Who are the best people, the most competent people, the highest performing people to help your team succeed?

142 Team Change: Success Skills 3

Competent Team Members

Vital Behaviors for Success.

In the book, *Influencer*, the authors point out that great change agents—those who get the best out of people during challenging conditions—know how to define the Vital Behaviors for success. What are Vital Behaviors? Let's call them the "critical few" behaviors required for breakthrough performance; for ultimate success; for real change to happen.

Team Leaders can also benefit from defining the Vital Behaviors for team success. Here are the questions.

What Are The Critical Technical Skills For Team Success?

What Are The Critical People Skills For Team Success?

Key word: critical. These are not "nice to have" skills. They are "need to have" skills. Actually, let's expand the word "skills" to include, knowledge, skills, abilities and characteristics.

What are the top 5 skills needed for your team's success?

YOUR NOTES:

143 Team Change: Success Skills 3

Competent Team Members

Balance

A key to success in much of life is: BALANCE!

And highly productive teams are no exception. In addition to having the right people with the right skills, teams need the right BALANCE of those skills. For example, a team leader may need to ask, "How many ________ people do we need?" Fill in the blank with descriptors such as these…

- "Nuts and Bolts"
- Idea
- Conceptual
- Action-Oriented Doers
- Negotiator-Type
- Planners
- "Big Picture"
- Detail-Oriented
- Meticulous
- Conscientious
- Inspiring
- Persuasive

- Technical
- Analytical
- Or any other descriptor you choose.

What balance of skills will lead to team success?

144 Team Change: Success Skills 3

Competent Team Members

Technical and People Skills Required.

"She's a technical star, but not good with people."

~Countless American Managers

Teams must work with TASKS and PEOPLE. Therefore, teams need people with:

TECHNICAL SKILLS: knowledge, skills and abilities to complete the appropriate tasks.

PEOPLE SKILLS: qualities, skills and abilities to resolve issues.

Some experts expand on this idea by creating roles for teams, some roles from each category:

Task Focused Behaviors: behaviors that assist with the accomplishment of the task or assists with the process.

- **Initiator**: proposes, suggests, defines
- **Informer**: offers facts, expresses feelings, gives opinions
- **Clarifier**: interprets, defines, clarifies
- **Summarizer**: connects, restates, concludes
- **Reality Tester**: offers critical analysis

Maintenance Behaviors: behaviors that focus on the members of the group.

- **Harmonizer**: reduce tension and reconciles disagreements
- **Gate Keeper**: facilitates participation by all
- **Consensus Tester**: checks on decision making process
- **Encourager**: warm, responsive, shows acceptance
- **Compromiser**: admits error, avoids conflict

The roles do not need to be assigned to just one person each, but can be fulfilled by anyone on the team at any time.

Of course, some people's talents make them better suited to some roles more than others, but in the end the key is to make sure your team is balanced and that roles from both categories are active. Make sure to account for both task and people in your team.

How can our team balance task and people?

145 Team Change: Success Skills 3

Competent Team Members

Demonstrate A Strong Desire to Contribute.

Team Leaders agree. One competency teams need for success is:

A Strong Desire To Contribute.

So smart team leaders help team members see how performing on the team connects to their career aspirations, desires and development. In other words, smart team leaders help team members answer the WIIFM question,

"What's In It For Me?" To Be On This Team.

The clearer the common goal, the easier to make the connection. And, the higher likelihood that team members will be motivated to make their contributions. In the end, it's not enough to just show up for team meetings. Team members must constantly think,

What Is The Best Way For Me To Contribute?

And then deliver on the answers.

It takes inspiring leadership with an eye for what makes people tick and many deliberate reminders to reach for excellence to bring out the best in people like that. WIIFM is one tool that can help. A vision is another. Constant and candid communication helps too.

ON A SCALE OF 1-10 (1=low, 10=high)

How motivated are you to contribute to the team?

What is the best way for you to contribute to this team?

146 Team Change: Success Skills 3

Competent Team Members

Work Well Together.

Ask any team leader what it means to be a team and many would answer, "Working well together." The answer implies that good team members are also good "people-people." That is, they possess and use strong people skills. Knowing how important people skills are to working better together, aka teamwork, team leaders choose team member with the right people skills right from the start. And they provide training to sharpen those skills as the team work continues. Working together well is:

- Building on the best ideas and strengths of self and others.
- Looking beyond positions and identifying underlying interests of each team member and stakeholders.
- Sharing views openly, honestly, clearly, and respectfully.
- Listening with an open mind, seeking to understand before seeking to be understood.
- Probing for deeper understanding.
- Discovering points of agreement as well as expressing disagreement.
- Balancing self-interest with others' interests and always putting the team goal first.

One final note, team leaders agree that even ONE team member without the ability to work well with others, can dramatically affect the team—negatively. One disruptor, rabble-rouser, naysayer, slacker or prima donna can ruin the entire team. Even top performers ramp down performance in the presence of more negative players like these.

Disruptors: (or Interrupters) behaviors that get the group off-task or make the process difficult; it may be behaviors that are directed to a particular individual within the group.

Aggressor: deflates others, attacks ideas

Blocker: disagrees & opposes beyond reason

Dominator: asserts superiority to manipulate, controls

Cavalier: plays more than works; generally light-hearted and non-productive; does not take tasks seriously

Avoidance: pursues special interests to avoid task

How well does your team work together?

How effective are team players with people skills?

147 Team Change: Success Skills 4

Complete Commitment

What is commitment?

dedication, energy, devotion, loyalty, enthusiasm, willingness, identification, intensity, effort, investment, drive, allegiance, faithfulness, responsibility, obligation, duty, pledge, agreement

These words describe complete commitment. Teams with complete commitment from team members enjoy incredible dedication to the common goal along with serious investment of energy along each step of the way in the journey to fulfillment of that goal. The team works hard. Very hard.

In addition, each team member puts the common goal ahead of his or her self-interest. The result? The team takes on a life as an entity in its own right. A team identity is born. And along with the team's identity, team members begin to make the team's identity their own as well.

No, individuals on teams don't lose their individual identities. But they identify with team first, self second in the same way that they put team goal first and personal goals second.

This phenomenon is illustrated in a very exciting and memorable way in the film *Miracle*. The film is a docudrama about the United States men's hockey team, led by head coach

Herb Brooks. The team won the gold medal in the 1980 Olympics, against tough odds. No one really believed the US team could win the gold. The USA team's victory over the heavily favored Soviet team in the medal round was dubbed the Miracle on Ice and paved the way to victory with gold.

Herb Brooks was a remarkable coach, with a deep understanding of the talents of the players he chose. From the start, Brooks chose players that were not thought of as stars and therefore were not considered the right picks by many hockey experts. Brooks' goal in picking players, however, was to create the kind of extraordinary teamwork necessary to win Olympic Gold. He wanted a blend of talents and skills that would result in great team play.

Early in training, Brooks drove players hard to develop skill and endurance, as one might expect any fine coach would do (although Brooks pushed harder than most and to a level that his assistant coaches balked at). In addition, however, Brooks created a ritual, regularly asking individual players for their identities. Early on, players answered by citing the past team played for. University of Minnesota. Boston. When the team finally started to meld a bit later in the film, players would answer, "USA."

Players on the Miracle hockey team moved from "me" identity to "we" identity, creating the powerful synergy in teamwork that propelled them far past the mere collection of individual potentials. Working better together made the USA Olympic team far, far more than a mere sum of their parts.

What is your team's level of complete commitment?

YOUR NOTES:

148 Team Change: Success Skills 4

Complete Commitment

Increased Involvement = Increased Commitment

Try this formula on for size. Increased involvement = Increased commitment. Yes. It's that simple. If you want a committed team, get the players involved. Especially, involve them in:

PART ONE: Common Goal

PART TWO: Plans to achieve Common Goal

PART ONE: Pitch the goal with benefits—to the individual and to the company. Communicate the goal in crystal clear terms. Facilitate full discussion about the goal. Probe for everyone's views. Agreement? Disagreement? Possibilities? Impossibilities? Drivers of success? Roadblocks? Discuss it all. Then discuss it again.

PART TWO: Craft the plan to achieve the common goal. Get everyone involved from the beginning. It must be the team's plan, not just the leader's plan or the company's plan. Hear the voices of every team member to ensure what experts call "buy-in." That is, when people "buy-in"—when they believe and commit (even if they don't fully agree)—those same people will work harder to make things happen.

How involved are the team members on your team?

149 Team Change: Success Skills 4

Complete Commitment

There is no "I" in Team!

This well-worn phrase is a bit over-used and over-stated, but the underlying point is still a good one. Winning teams demand unity. Team first. Priority 1. Self second. And yet, self must still exist.

A look at the extremes might help clear things up. Too much unity can lead to what is known as Group Think. That is, people leave their individual critical thinking behind in favor of simply endorsing the majority view, even when those same people sometimes silently disagree. Too much unity can also lead to blind conformity, a first cousin to Group Think. Neither Group Think or blind conformity are good for teamwork.

Let's contrast too much unity with too little unity. Teams with too little unity suffer from endless, destructive conflict and paralysis by analysis. In addition, as one might guess, such teams don't achieve what they set out to achieve. And they rarely embrace the kind of commitment that leads to holding people accountable for team contributions. Mediocrity abounds. Bickering reigns. In the end, too little unity is a case of too many selves and too little team.

So, there's personal identity and team identity. Great teams balance both. How? Here's one piece of sage advice offered in the book *Teams: What Must Go Right/What Can Go Wrong*.

"Put the team's goal first. Personal interests must be second. And ask every team member to do the same. Members who do not follow this rule and get carried away with their personal agendas and ambitions must get on board or leave."

How balanced are your team's individual identities and team identity?

150 Team Change: Success Skills 5

Collaboration And Cooperation

Trust Drives Teamwork.

When people talk "team," what they most often mean is collaboration and cooperation. And collaboration and cooperation are all about trust. Trust and teamwork. They're inter-twined. You can't have one without the other. Teams with high trust:

1. **Focus on the goal**. In the absence of trust, people are guarded. Personal agendas trump the team's goal.

2. **Communicate and coordinate.** They confront issues candidly, deal with things head-on, and keep each other informed.

3. **Embrace risk-taking.** Difficulties are openly discussed. Dissenting views are welcome. Failures are carefully examined in order to learn lessons for next time. Problems are dealt with, not hidden. Action trumps caution.

4. **Pitch in to help.** Teamwork experts call this "compensating" which means that when one team member falters, another steps in to pick up the slack. Without trust, such compensation is not likely to happen. With high trust, it happens with ease and without hesitation.

How well does your team collaborate and cooperate?

151 Team Change: Success Skills 5

Collaboration And Cooperation

Get People Involved.

If you want people to collaborate (pool their resources to achieve the goal) and cooperate (work together to get the job done) give them many ways to be involved and make choices.

Involvement: All opinions are surfaced; everyone has a say; reservations are discussed; no topic is avoided; everyone participates in planning (if not as a full contributor from the beginning, then in discussion later).

Autonomy: People enjoy the freedom of choice; no one works under dictatorship, demand or coercion; instead, people work under strong leadership, influence, and high expectations. To the highest degree possible, team members decide the proper course of action for the team.

Research verifies this. Both must exist.

How do you involve team members?

How do you give team members autonomy?

YOUR NOTES:

152 Team Change: Success Skills 6

Challenge

Great Teams Never Settle.

Great teams never settle. No matter how much they have achieved or how good they are, great teams never become complacent. They raise the bar and keep raising it—otherwise known as continuous improvement.

So, today's success is celebrated. But it's never enough. Great teams are determined to go to the next level.

They set a new goal and start again.

How do you raise the bar on team performance?

YOUR NOTES:

153 Team Change: Success Skills 6

Challenge

How do great teams raise the bar?

Great teams insist on performing to high standards. The key words are "insist" and "high." Doing well most of the time? Well, that's just not good enough. And average performance? Well, that's also not good enough. The best teams perform to high standards and seek to perform beyond those high standards every day and in every way.

Standard: The word means pressure to perform at a challenging and agreed upon level. Here are some questions that help define standards.

- How skilled must our players be?
- How much initiative and drive must we have?
- How must we treat one another? What is the code of conduct?
- How much discipline must we exercise?
- What are the rewards of success?
- What are the consequences of failure?

What are the performance standards for your team?

Are they high enough?

154 Team Change: Success Skills 6

Challenge

Strive for Excellence.

Do you want a team that strives for excellence? Here are 5 ways to do that.

1. Pick team members who hold high personal standards.

2. Encourage team members to hold one another accountable.

3. Keep the rewards of success and the consequences of failure in mind—all the time.

4. Leverage the external pressures to succeed—top management and the media come to mind.

5. Select a team leader who models and expects excellence.

The research on high standards and teamwork says that influences from outside propel teams to higher standards or otherwise. What others expect impacts the level of achievement teams work toward. This is a sort of self-fulfilling prophecy.

So, please know and discuss how outside influences impact your team. Evaluate whether expectations from outside will boost your team's performance or potentially hinder it.

Engineer your team's performance accordingly. If the forces are in the team's favor, fine. Keep examining whether that remains the case. If the forces are against the team, plan for ways to overcome popular opinion.

History is full of examples of those who "overcame" the expectations of others, like the USA 1980 Olympic Hockey Team. It can be done. But deliberate discussion can help to do that.

How does your team accomplish the 5 keys to excellence?

155 Team Change: Success Skills 6

Challenge

The Devil Is In The Details

"...people with high standards are those people who do ordinary things in an extraordinary way."

~Anthony Rucci, Team Leader, Baxter Travenol/American Hospital Supply merger

Anthony Rucci was talking about a woman—a mere clerical worker with presumably humble assignments—who did those assignments well above and beyond what was expected. Clerical or not, humble position aside, she dramatically and positively influenced her team.

Or, consider the Blue Angels—the Navy's very famous and very successful flying team. The Blue Angels aren't just precise while performing, flying through the atmosphere to the awe of the crowds.

After years of being in the Navy and being together, the Blue Angels still practice the march to and from the field. Why? Adherence to a standard of excellence in the small things reinforces achieving excellence in the big stuff.

Great teams know the devil is in the details. Excellence is not just a one-time event or a sometimes deal. Excellence is a way of working—a discipline—in all things. Large and small. Visible and behind-the-scenes.

ON A SCALE OF 1-10 (1=low, 10=high)

What is the level of excellence of this team?

What is the evidence?

156 Team Change: Success Skills 6

Challenge

A team's performance standards must not only be high...

They must also be clear and concrete. They must be easily observed, measured and communicated. So, look at these examples:

good attitude, very committed, work hard, be proactive, be persistent

Those phrases don't say much. Compare them with these statements:

- "Focus on what can be done, not on what can't be done."
- "Stay with the team until all the work is done each day."
- "Do everything that is required and add one more thing—that wasn't asked of you."
- "Step up and volunteer to help. Don't wait to be asked."
- "Don't let the first failure stop you. Or the second. Or the third. Get help. Keep with it until it the job gets done. And done right."

Better? Clearer? Easier to see? The lesson is: set the standards using clear, concrete words.

How can we state clear and concrete standards?

157 Team Change: Success Skills 6

Challenge

Yesterday's Success Is Today's Opportunity.

Picture this. The Blue Angels—post performance. Each Blue Angel is seated around a table. Most of the participants actually flew during performance. A few others were observers on the ground. All are prepared to critique what happened. They will take about twice as long to debrief as they spent performing.

Ask a Blue Angel and you'll find out that during debrief, rank and experience just don't matter much. Even the Boss, as they call their leader, is not Boss anymore. He is equal to everyone else.

During debrief, everyone must be wide open to criticism. No holds barred. The very survival of the team depends on it.

If you were privileged enough to be present at debrief, what might you hear? Would you hear the Boss critiquing team performance? No. You would hear each Blue Angel critique his own performance, owning up to mistakes, no matter how small.

The Blue Angels call mistakes "safeties" and follow each revelation of a "safety" with the ritual words, "I will fix it." The Blue Angels epitomize accountability like no other I have ever witnessed and their methods for evoking excellence prove it.

Like all top performers, for whom excellence is a life blood, not just a wish, critiquing performance—including giving and receiving candid feedback—functions like red blood cells, carrying oxygen to fuel everything else good in the system.

Why all this focus on what went wrong during a Blue Angels performance? And why do this after every performance, no exceptions? Simple. Discussing what went wrong reveals what could go right. Focusing on what went wrong lights a way to what could be improved—perfected—next time around. This is a never-ending, no exceptions process which reminds that excellence is a way of life, not just a once in a while thing.

How does your team critique performance?
When? Where?

What is the process?

158 Team Change: Success Skills 6

Challenge

Let's go back to our Blue Angels during debrief.

We've talked about what you WILL see. Picture Blue Angels around a table, each Angel sharing his performance mistakes and each Angel vowing to fix those mistakes. You would also see other Blue Angels offering feedback about the performance. This is to ensure that no Blue Angels pilot suffers from "blind spots," that is, flaws in performance which he cannot see.

But what WON'T you see during a Blue Angels debrief? DEFENSIVENESS.

The Blue Angels get it. They know:

Looking At Many Flaws Provides Just As Many
Opportunities To Improve!

Period. Yes, it's very powerful to know one's strengths and to develop those as a way to excellence, high performance and success. It's equally important to know one's weaknesses and get on with the business of improving those. Why defend? There's not much point. Defending one's current position is like saying that you're quite satisfied with things JUST THE WAY THEY ARE. Where's the opportunity in that?

How well do you receive feedback?

Do you remain open and non-defensive?

Do you listen and learn?

Do you probe to deepen your understanding?

159 Team Change: Success Skills 7

Leadership

What must team leaders do?

Here's a list of team leader duties.

1. Keep the focus on the team's common goal and off of competing issues.
2. Demonstrate the utmost team commitment; be a role model for the team.
3. Protect the team from too many competing priorities. Maintain focus.
4. Be fair and consistent with team members.
5. Coach for the highest performance. Confront performance problems.
6. Encourage open team discussion—information and idea sharing.

How well does your team leader fulfill his or her duties?

YOUR NOTES:

160 Team Change: Success Skills 7

Leadership

What must team members do?

How should the team treat one another? Here are the duties of team members.

1. Understand his or her role and responsibilities.
2. Make decisions based on facts first, opinions second.
3. Collaborate with other team members. Believe that all of us is better than one of us.
4. Put the team first, self second.
5. Do whatever it takes to achieve the common goal.
6. Communicate openly and candidly—information, perceptions, and feedback.
7. Offer help to other team members as needed.
8. Do the best every day.
9. Stand behind team decisions.
10. Engage in respectful conflict for the sake of excellence and success.

11. Listen and learn from feedback. Don't defend.

What is your greatest strength as a team member?

What will you improve for the sake of the team?

161 Team Change: Success Skills 7

Leadership

Empowerment: Not Just A Word For Teams

For great teams, empowerment must be more than a word. It must be a reality. Teams thrive under the following empowering conditions. And, a great team leader makes sure these conditions exist.

1. Get challenging assignments and meaningful responsibility.
2. Room for initiative and choice to achieve results.
3. Team members get more challenging assignments in order to develop and grow.
4. Recognition and rewards for superior performance
5. A team leader who stands behind the team.

How empowered is your team?

YOUR NOTES:

QUOTE THE WISE - TEAMS

In this section, find more quotes, as in an earlier section, but this time around, the quotes are specifically designed to inspire teamwork.

YOUR NOTES:

162 Chinese Proverb

Ants can move even a mighty mountain.

~ Chinese Proverb

Teamwork rocks. It does. It may not be easy to form a real team. But once you have done so, expect great results.

Start with the most important ingredient. Offer a clear, compelling goal for the change. Rally the troops around that goal.

After that, expect the team to have a healthy self-interest, but expect each and every team member to put the big team goal ahead of individual self-interest.

What is the team's vision?

Has the entire team committed to the vision?

163 Eleanor Roosevelt

The future belongs to those who believe in the beauty of their dreams.

~ Eleanor Roosevelt

Most of us live sensible lives. We do the right things, at least most of the time. We follow the rules. We faithfully go to work and do our jobs. We take care of our families and friends. Good!

The downside? Sometimes we can become so wrapped up in doing the daily chores that we forget to dream. And we need to dream. Dreams are a source of life's inspiration. No wonder then, when the organization has a dream, we shrink back with caution. We've forgotten the power of dreams.

What are some dreams you have forgotten or have put on the back burner?

164 Leo Burnett

When you reach for the stars, you may not quite get one, but you won't come up with a handful of mud either.

~ Leo Burnett

Progress can be frightening. Success is not guaranteed. Deep down—or maybe not so deep down—we know that. At the same time, life and work without "reaching for the stars" now and again, takes us nowhere new. That IS guaranteed.

So reach for the stars. Do so intelligently, but do so. No, you are not guaranteed success. Okay. You are guaranteed to learn, grow, and flex your change muscles.

What stars can you reach for?

YOUR NOTES:

165 Anonymous

Inch by inch, it's a cinch. But by the yard, it's very hard.

~ Anonymous

During change—any change—shrink the size of the steps you take. Start with the Change Vision and make the Change Plan. Take the Change Plan and make some Change Steps. Chunk the Change Steps down to even smaller steps. Making steps small makes them easier to do. Finally, do

One Small Step At A Time.

A psychologist worked with a woman who was so depressed, she just could not start exercising, even though she knew she "should." The psychologist, knowing the power of small steps, recommended that the woman start by standing on her treadmill for two minutes each day. Even a small step can start the momentum toward success flowing.

The Japanese, who make kaizen so famous, applaud the power of what they call "little ups," "little ups" every day. They founded an entire way of doing business on the practice of continuous, small improvements, which—all added together—moved a whole economy post World War II.

Achieve the Small Wins. See the Small Wins. Celebrate the Small Wins. Day by day.

What small steps of this change can you achieve?

166 Anonymous

Courage is simply the willingness to be afraid and act anyway.

~ Anonymous

Many people assume that those who are successful, those who tackle the biggest challenges, and those who pursue the seemingly impossible dream are those who do not feel fear. The assumption is wrong.

Everyone fears something and most of us possess the most typical of fears like:

- Fear of failure
- Fear of embarrassment
- Fear of being wrong
- Fear of conflict
- Fear of loss of status
- Fear of loss of power
- Fear of success
- Fear of public speaking

- Fear of learning
- And on and on.

The secret of the successful is not that they don't feel fear. The secret of the successful is that they feel their fears and move ahead anyway.

So, tune into what you are afraid of. Face those fears. Then, just as you would do for a good friend, coach yourself kindly but firmly forward.

Feel the fear and move ahead anyway.

What are your fears during change?

How can you coach yourself through those fears?

167 Dr. Seuss

You will come to a place where the streets are not marked. Some windows are lighted. But mostly they're darked. A place you could sprain both your elbow and chin! Do you dare to stay out? Do you dare to go in? How much can you lose? How much can you win?

~ Dr. Seuss, *Oh, The Places You'll Go!*

Dr. Seuss delivers a lot of wisdom. Sometimes the way is uncertain…that's the only thing we can know for sure. Is the reward of taking the next steps to fulfilling your dreams worth the risk? Only you can choose.

Have you weighed the pros and cons of this change?

168 Charles E. Rice

Think of managing change as an adventure. It tests your skills and abilities. It brings forth talent that may have been dormant. Change is also a training ground for leadership. When we think of leaders, we remember times of change, innovation and conflict. Leadership is often about shaping a new way of life. To do that, you must advance change, take risks and accept responsibility for making change happen.

~ Charles E. Rice

Change as an adventure and a training ground. Marvelous! How much better to see change in these ways as compared with seeing change as

A headache. Distracting. Frightening. Unnecessary. Stressful.

Choose your view. Change your change experience.

How do you choose to view change?

How can you change your mind if need be?

169 Essays

As the births of living creatures are at first ill-shapen, so are all innovations at the births of time.

~ Essays

Are there any guarantees during change? We can think of one. There will be bumps in the road. Trouble, challenges, and bloopers lie ahead. Not everything will go as planned.

Expect the unexpected and you'll feel better about it when it happens. Expect perfection and you will certainly be disappointed.

But just as babies become even more perfect and lovely as they grow, so do many changes.

What bumps in the road should you expect during this change?

170 John Kenneth Galbraith

In all life one should comfort the afflicted, but verily, also, one should afflict the comfortable, and especially when they are comfortably, contentedly, even happily wrong.

~ John Kenneth Galbraith

Some thrive on the challenge of constant change. Most, however, need some prodding to get out of their comfort zones. The prodding can come in a somewhat urgent fashion, the result of the act of taking a hard look at current weaknesses or failures. Or, the prodding can come enthusiastically, the result of a vision of the possible. Or, the prodding can be more than a prod—it can strike and strike hard—as happens with crisis.

What is prodding you out of your comfort zone?

What is prodding your team or organization out of comfort zones?

171 Unknown

Change is the essence of life. Be willing to surrender what you are for what you could become.

~ *Unknown*

Knowing oneself and taking pride in oneself—as an individual, team, or organization—is healthy, up to a point. That same sense of identity and pride, however, can keep us focused too much on today and not enough on the possibilities of tomorrow. We can tend to get too comfortable. We can start to become lax and stagnant, failing to recognize the vibrant possibilities that life and work can offer those who will see them. In short, over time, swaddled in the warm blanket of our comfort, we fail to grow.

To become all that we can be, we must give up some of today's pride, comfort, security for tomorrow's potential.

What must you surrender today to reach your potential tomorrow?

172 Pablo Picasso

I am always doing that which I cannot do, in order that I may learn how to do it.

~ Pablo Picasso

Picasso. He was that oh-so-famous artist who made the oh-so-funny-looking pictures. And what genius he was!

We think it's safe to say that Picasso lived "outside the box." And from the quote we know that he valued doing the difficult. The "never done before." The unknown.

And, he did so specifically in order to learn. Let Picasso's art at living life inspire you as much as his paintings inspire millions. Follow his example

What is something you want to learn?

173 Peter Drucker

Unless commitment is made, there are only promises and hopes…but no plans.

~ Peter Drucker

Teach people to do frequent Commitment Checks during change. Ask for honesty as they rate their individual commitment to the team's work. The goal of this exercise? Full disclosure of candid views related to the team's progress. NOT the watered down or sugar coated version. The truth, stated in a way that can help the team succeed.

Practice will help you and your team perfect this exercise. Do the exercise regularly, every few weeks or so, depending on the complexity of the team's goal and project. Start by posing the questions below. And, offer an answer of your own as well.

On a scale of 1-10, 1=no commitment and 10=full-fledged commitment

What is your current level of commitment to this team's ultimate goal?

Why?

174 Unknown

Teamwork divides the task and multiplies the success.

~Unknown

Change efforts thrive when teamwork prevails. During any change initiative—large or small—the fulfillment of the change vision happens faster, easier, and more efficiently when the team pulls together. Each carries his or her part of the load while staying aligned to the common goal.

True teamwork can be challenging to create, but the multiplying effect on success makes it well worth it.

What can you do to increase the level of teamwork in your organization?

175 Michael Jordan

Talent wins games, but teamwork and intelligence wins championships.

~Michael Jordan

Who better than Michael Jordan to remind us? Talent is great. But, without teamwork and intelligent play, talent may come up short. If so in sports, perhaps it's the same in business.

Recruit the best talent. Develop them well. But, to seal the deal on change success, create strong teamwork to execute the most intelligent plan possible. Then, let them all go to play their hearts out!

How can you ensure you have recruited the best talent for your team?

YOUR NOTES:

CONCLUSION

We hope you found this journey as inspiring as we did. Change is challenging for everyone: for those who shy from it as well as for those who embrace change whenever possible. The important thing to remember is that changed is a mixed bag.

Yes, change can bring uncertainty, the unknown, occasional stormy waters, a failure here and there, and the anxiety to go with all of it. At the same time, change can also enrich, enlighten, strengthen, challenge, awaken, and open up possibilities we never dreamed of.

We hope you will realize the former, without denial. We hope with equal fervor that you will squeeze all the zest you can out of the latter.

We wish you the best of luck with your change process.

Chris and Deb

YOUR NOTES:

THE CHALLENGE OF CHANGE

More than one expert has said that the only constant in life is change. The statement—stated so often that it's become cliché—is true nonetheless. Even those who don't want to change—and there are plenty—find themselves forced out of complacency, like it or not. Here's why.

There are external pressures.

- Local competition
- Global competition
- Financial pressures
- Technological advances
- Customer and shareholder expectations
- Political and economic uncertainty
- Government regulation
- And on, and on, and on…

There are internal pressures.

- Top Management and Board expectations
- Management expectations all around, at all levels
- Competing political agendas
- Employees, from disgruntled to fearful to sabotaging
- Silos in conflict

- Strategic goals
- Short-term goals
- Mounting deadlines and demands

With stressors like these, it's inevitable. Companies that want to survive must change. And companies who want to do more than survive—those who want to thrive—must change even more. Somehow. Some way. Wise leaders know it's so.

Change Efforts Can Fail

Yet ironically, management gurus who study and measure such things tell us that most formal organizational change efforts fail, in at least one significant aspect. Some say the failure rate is as high as 85%. In the real world it looks like this. A company gets in a game of change with the highest aspirations and most promising vision only to throw in the towel after just one inning. Score? Progress 0. Cynicism 5.

It's no wonder that so many employees complain about the "flavor of the month" initiatives in their companies. Given the dismal statistics about the success of change, they're right. Consider just two of the most common types of initiatives that fall into the change category—process improvement and technological advances. How often do those initiatives fail to produce desired results? Estimates range from 60-70% of the time. Not so good.

But there IS something good. When a company DOES achieve successful change, the results typically stand out. They're startling. And profitable. Consider Jack Welch, former CEO of GE. He led GE through successful organizational culture shift, now heralded in many articles and books. What's all the fuss about? Well, in the case of GE, some powerful and compelling words come to mind. Innovation. Agility. Performance. Extraordinary leadership. Over the span of Welch's tenure at GE, his efforts at leading change ultimately resulted in profits. Big profits.

Looking at the astonishing results at GE (and others like the IBM turnaround and the Virginia Mason transformation, to name just two), it's no surprise that other top management teams have wanted and still want to get into the change game. So, how to set the stage for extraordinary change? How to navigate the churning waters of the change journey? How to keep from throwing in the towel before the big win? In short, how to transform change into real competitive advantage?

Change experts offer lots of answers to those questions. And WOW! Change Cards dispenses those answers and other expert advice in small, easy-to-digest bites. We cannot take credit for the ideas. We can take credit for packaging them in a way that is easy to sort, digest, and do.

Change Is Just A Card Away

Well, more realistically, change is many WOW! Change Cards away. But WOW! Change Cards are an easy-to-use tool to support you through change. WOW! Change Cards will give you a change education. Offer solid change advice. Lift sagging change spirits. Stimulate some robust change conversations.

WOW! Change Cards can be used by:

- Individuals
- Facilitators
- Trainers
- Executives
- Business owners
- Managers
- Supervisors
- Departments
- Teams
- Leaders of all kinds

In so many ways:

- Self-directed learning
- Self-reflection
- Classroom training
- Staff meetings
- Retreats
- Department bulletin boards
- Brainstorming sessions
- Planning meetings
- Team building sessions
- Project management

- Progress meetings
- Settings of many types

In order to:

- Study change concepts, techniques, and ideas—at your own pace, one small bite at a time.
- Equip executives, managers, supervisors and employees at all levels with change advice, on the spot.
- Stimulate group discussion about important change questions and issues.
- Infuse spirit and fun into even the most serious business discussions.
- Reinforce classroom training change lessons.
- Get everyone—not just a dominant few—involved in change discussions.
- Help people think "outside the box."
- Support people through the stress of change.
- Fulfill the adult learning needs for auditory, visual, and kinesthetic learning methods.
- Keep meetings, training, and other sessions lively and fast paced while focusing on the topic of change.
- Take change lessons and inspiration "to go." Review them anytime and anyplace.
- Display one change lesson per day. On your desk, or a table, or a mirror. Anywhere.
- Manage the stress of change in your life and work.

YOUR NOTES:

HOW ADULTS LEARN

You work with adults. And effective change depends on learning. Lots of learning. So, it pays to know how adults learn. Here are the principles that—when skillfully applied—the learning gurus say create strong learning environments and experiences for adults.

To begin, realize that adult learners have special needs. Adults receive information from each of the five senses. Take a look:

- 83% of information from what we See
- 11% of information from what we Hear
- 1% of information from what we Taste
- 1.5% of information from what we Touch
- 3.5% of information from what we Smell

Get the picture? TELLING people what to do isn't effective. SHOWING them what to do works much better. (Are you surprised that "smell" outdid "taste" and "touch"?) And, to teach so it sticks, use all of the senses. In addition, apply these adult learning principles:

Adult Learning Principles:

Reinforce It!

Adults do what is rewarded. In training, give lots of chances for people to answer questions and then reward their answers. Even an affirmative, encouraging look is a reward worth offering. Or, pass out token-type rewards like pieces of candy.

Emotions Rule

Boredom is out. When their eyes glaze over, liven things up. Tell a story. Ask them to tell stories. Share a joke. Play a game. These methods make it easy to learn new information and anchor the information with the power of emotion. Voila! Learning that sticks.

Be Active, Not Passive

Make training active, not passive. Ask questions. Encourage them to ask questions. Answer questions. Question their answers. In short, training—like great communication—is a meeting of the minds. It's best when it goes both ways. The good old-fashioned information dump just doesn't work.

Practice, Practice, Practice

Top performers get to the top by lots of practice and feedback. Ask participants to talk about what they will do with what they've learned. Ask them to create action plans for application following training. Give them real-world scenarios to role play. Conduct fishbowl sessions so the larger class can comment on the practice efforts of a smaller part of the class. Create simulations that test principles and practices acquired during class or e-learning.

Link the Old with the New

Find out what the participants already know that can help them learn new material. Mine participants' experiences by asking questions and listening carefully to answers. Ask questions that elicit full descriptions, similar to behavior-based interview questions. For example, ask "Tell us about a time when you were treated differently than everyone else in a group setting? How did it feel?" Or, "Everyone faces significant change sometime during life or work. Tell us about a time when you faced significant change. What happened? How did you handle it? What were the results?" Then, link past experiences with new material, making it easier to digest, understand and stick.

Different Strokes for Different Folks

Different people learn in different ways. Smart trainers know this and plan learning experiences that appeal to different learning styles. For example, ask participants to write, talk, draw, and act. Ask them to learn individually, in pairs, in teams and online. Shake things up with a variety of ways and means for learning.

Get Real!

Adults prefer to learn what they NEED to know as opposed what is NICE to know. So, provide less theory and more nuts and bolts. Related to that, adults thrive on learning the knowledge, skills and abilities necessary to do the job tomorrow, or at least soon thereafter. Give participants the important skills and let them practice, so they're prepared to go out and do.

Some More Specific Tips

Those are the main adult learning ideas to consider. Now, for a few more specific ideas...

Keep it practical and help people solve real problems.

- Use examples and stories to make concepts real, tangible, and practical.
- Encourage discussion about how learning will be used.
- Enhance learning through collaboration. Plan for lots of talk with peers.
- Discuss how the learning may be hard to apply. Ask, "What are the barriers and potential stumbling blocks to using this material?"
- Ask, "What are some problems this material will help you solve?"

Catch them doing things right, as often as you can. Prepare learners for success.

Provide ways to practice new skills and discuss practice results throughout the learning experience.

- Help learners become more and more successful, one step at a time.
- Give clear expectations for each practice round to boost confidence.

- Reward and recognize successes, small, large and in-between.
- Ask, "How can we help you succeed?"

Go back to go forward.

- Ask participants to share past experiences that may be relevant to new learning.
- Pinpoint participants' learning needs. Ask for input about the agenda.
- Adjust the timeframes on the agenda to fit participants' needs.
- Ask participants what they really need to know about a topic.
- Quickly adjust to learners' needs by being a flexible trainer.
- Design follow-up support for learners as they apply new knowledge and skills; i.e. a learning help line.
- Ask, "How can we best support your ability to apply what you've learned to the job?"

Establish a safe and respectful learning environment.

- Make learners physically comfortable during the learning experience.
- Be well organized and make good use of learning time.
- Avoid big words and jargon. Don't "talk down" to participants.
- Validate learners' views even when offering a different

perspective.

- During class, frequently ask for feedback about how the learning is going.
- Get input to enrich the learning experience.
- Ask, "How can we make this learning experience the best you've ever had?"

Provide choice and alternatives to fully engage the adult learner.

- Show learners the gap between what they know today and what they need to know tomorrow.
- Share your agenda and ask for reactions and input.
- Discover what the learners know about the topic already.
- Ask what they would like to know about the topic.
- Create a learning plan with a variety of learning methods.
- Incorporate action-planning time into the learning agenda.
- Ask, "What will you do with this when you get back to work?"

Skills For Facilitating Learning

In the old days, managers gave answers. Workers used those answers. Times have changed. These days, the smartest managers are teachers. They want the employees to have the answers or to discover their own answers. So, those managers facilitate learning instead of dispensing answers like pills from a bottle. The old job description of "manager as know-it-all answer provider" no longer applies.

WOW! Change Cards offer a juicy opportunity to facilitate learning, helping people find their own answers, instead of just dispensing them. And during change, peoples' ability to learn is a key ingredient to success. It doesn't do much good to order people to "CHANGE ALREADY!" It's much better to engage in training—both formal and informal—about why the change is necessary, what might be achieved, what problems might occur, how things will be tackled and what everyone stands to gain.

WOW! Change Cards are structured to support on-going, day-to-day learning. In addition to some content, most cards have a question or two to get people thinking and talking. By using the cards—with the questions—executives, managers, supervisors, project managers and team leaders can easily facilitate learning. And employees can take charge of discovering the best answers.

Here are some methods for facilitating learning. There are also learning methods on some actual WOW! Change Cards, but the ones included here build on those ideas. Use any of the methods, whether you are a bona fide trainer or a change leader who is taking your new role as facilitator of learning to heart. Use one. Use several. Over time, you may use all the methods here. Enjoy!

Get Them Involved And Keep Them Involved

Ask for volunteers: to present, to report, to ask a question, to pose a challenge, to keep track of time.

Make a job jar. Fill the jar with roles that must be fulfilled during the meeting or training. Ask people to draw a job from the jar and do that job for the day. Rotate jobs at each session. Job examples? Time keeper, devil's advocate, summarizer, recorder, comedian, story-teller, interviewer, and the like.

Get frequent feedback. About the subject, issues, opinions or anything else related to the topic at hand. Ask for a simple "thumbs up" or "thumbs down" on a topic, idea or opinion. Or, pass out sign cards—"+," "-," "?"—to get instant feedback. Use symbols to get the kind of quick, yet accurate feedback you want. These days, you can achieve the high-tech version of gathering feedback using Twitter and PowerPoint technology.

Post a flip chart for feedback. Tell participants to write comments on the page throughout the meeting or training. Comments might include what's liked, what's disliked, ideas for follow-up, questions and so on.

Pose pre-meeting or pre-training questions and ask participants to come prepared with answers.

Ask participants to come prepared to share real-world situations that they would like to have addressed.

Use stories and quotes to start sessions with a bang. Ask each participant to share what a particular quote means to him or her. Ask each participant to share a story that relates to the subject at hand. Or, use a quote or story from a WOW! Change Card deck to start a lively conversation. (Stories are included in Volume 4-Inspiring.)

Assign a question and ask pairs to interview one another using that question. Ask for answers from each pair.

Give them journals. Encourage participants to take notes during each session. Ask them to emphasize action. What will they do as the result of the meeting or training?

Conduct a question and answer session. Ask participants to pose a question on an index card. Collect the cards. Shuffle and pose several questions to be answered during the session. Or, redistribute the cards and ask for answers in writing.

Do frequent round robin discussions. Pose an issue or question and go once around the group, getting an answer from each participant. Give the right to "pass" for those who do not have an answer they wish to share.

Brainstorm lists of all kinds. Some examples. What are the pros and cons of this idea? How could we improve this? What else do we need to consider? What are all the problems we must anticipate? Who do we need to include? What are the risks and rewards? How could we use this information? And on and on... Make the list long, including everything. Then prioritize.

Get physical. Get people moving by asking that they change seats. Change groups. Write comments on flip chart pages. Draw pictures on flip chart pages. Stand up to present. Raise their hands to vote. Any physical movement keeps things lively. Submit questions via online means.

Complete the sentence. Pose a sentence starter and ask for participants to fill in the blank. For example, "When I think about this change I feel..." "When I picture myself one year from now, I see..." "One thing I think we must do to succeed is..."

Ask the group to set ground rules for meeting behavior. This is a group conduct contract. Write the list of ideas on a flip chart. Post the list. Evaluate meetings or training based on the list. Ground rule examples: no interrupting, turn off cell phones, don't blame, do solve problems, all should participate.

Toss the ball or any other object (rubber chicken, Frisbee, stress ball) to facilitate participation throughout the entire group. One participant holds the ball and shares. Next, he or she should toss the ball to a new participant who continues the discussion.

Ask for ideas for agenda items. Use the ideas to craft the agenda for the next meeting or training session.

Line them up. To speak, that is. Pose an issue or question. Then, ask that all who wish to speak, to raise their hands. Assign a speaking order for all those who raised hands.

Call on people. If someone looks like he/she has something to say but is a bit reluctant to speak up, check out your perception.

Like this. "Hank, you look as if you have something to say. Is that right?" Do this with sensitivity. Respect it if someone answers, "No, not right now please."

Use the clock. A minute or two before break or ending, say this. "We have just a minute left and I want to make sure everyone has had a chance to speak. Susan and Sarah, we have not heard from either of you. Any comments?"

Start with some fun. Ask that participants raise their right arms—straight up in the air. Then instruct that on the count of 3, participants should point at any person in the group. The person with the most fingers pointed at him or her should start the discussion.

Argue both sides of an issue or question. Divide the group into two teams. Assign each team a position. One team should be for an issue. One against. Participants do not need to agree with the assigned position to fulfill it. Ask each side to prepare arguments. Then hear those arguments in a sort of trial by jury fashion.

Pass it on. Ask each participant to write a question or problem on a piece of paper. Then, ask participants to pass their paper to the person on their right. That person should write an answer/comment/solution in response to the question or problem. After a minute, ask participants to pass the papers on person to the right one more time. Repeat the process of passing/writing as many times as you wish before papers are returned to the original author.

Play ball. Ask participants to write an answer to a question or issue on a sheet of paper, crumble the paper into a ball, and toss the ball somewhere across the room. Once the paper balls are tossed, participants should pick up a ball, smooth it out, and read the response aloud.

Seek First To Understand.

Stephen Covey said, *"Seek first to understand before seeking to be understood."*

This is especially good advice for a facilitator. If you want to give great answers, sound advice, solid direction, discover the needs of others first. Or, put more simply, listen before you speak. And...

Ask Great Questions! Questions accomplish many things. They get people involved. They get people thinking. Questions keep people's attention. They show interest and care. Questions can also bring to light vital information, opinions and answers.

Here's a list of powerful questions. You'll find at least one or two great questions to ask at meetings or in training.

What's your reaction to this step of the plan?

What ideas do you have to improve this?

What other ideas do you have to improve this?

How do you know that actually happened?

How do you know that's a problem?

What's the evidence to support what is being said?

What are the circumstances surrounding this?

Who was involved?

What happened?

What led up to this?

If you could have anything, what would you wish for?

What is your worst fear?

What is the best thing that could happen?

What is the worst thing that could happen?

What thought process led you to that conclusion?

What is the root cause?

How would you like to handle this?

How should we proceed?

What part do you play in the problem? Others?

How can management help? Hurt?

What are the barriers to progress?

How can we deal with the barriers?

What else could get in the way?

What resources do we need?

What is the next best step?

What is working in our favor?

How will we know we're on track?

How will we measure success?

How will this look when we're finished?

What are all the gains we can count on when this is accomplished?

What is the return on investment?

How much can we lose?

If we do lose, how will we recover?

What can we learn from the situation?

How will we communicate our needs, desires?

How can we sell others on the idea?

How can we keep management posted on progress?

How will we keep each other posted on progress?

How can we break the tie?

How can we move through this disagreement?

What are our commonalities?

What are the real differences?

What are the possible compromises?

What are all the possible options to consider?

How should we choose the best one?

What are the criteria for evaluating the options?

What are the priorities to consider?

What order should we do these in?

What have we left out?

How should we celebrate?

The questions really are more important than the answers. The questions get people thinking in ways that answers cannot. Questions open up possibilities; answers narrow possibilities. Questions stimulate problem solving and innovation; answers prepare for action. Questions expand the range; answers focus it. Questions invite accountability from many; answers keep accountability with answer-giver. And , during change, perhaps the most important benefit to questions is: questions get people thinking in ways that naturally overcome resistance. By contrast, answers—especially those that sound like dictates—can actually arouse resistance.

Yes, at times you'll want to give answers and, depending on your role in the organization, may be required to do so. Just make sure that you've asked the questions whenever and wherever possible, especially if getting others to learn is your goal. In addition, right along with asking questions, make sure you've listened carefully to what others have to say. There's no better way to get lots of minds working on your organization's changes.

A Model For Learning

Learning is much more than the acquisition of knowledge. Excellent learning results in behavior change. So, any significant organizational change relies on the "learning as behavior change" formula. People must be equipped to do much more than understand the change. They must be able to execute it.

Here is a sequence for designing a learning that results in more than merely the acquisition of knowledge, and prepares people to skillfully act. Use it for any learning opportunity to include: classroom, meetings, coaching, groups, one-on-one. The same sequence applies in all cases.

1. Tell Them Why

Why learn this information and acquire these skills? What's the value? How do the information and skills apply to work? What are the benefits to the individual, team, organization? Simon Sinek talks about the power of why in his now-famous TED Talk, during which he cites Apple as being excellent at talking "Why," before "What" and "How." Sinek believes talking "Why before What and How" is a key to Apple's success in the marketplace.

2. Tell Them What

What are the learning objectives? What will they be able to think, say, and do when the learning is done? Be specific. List categories of information and types of skills, chunking the bigger stuff into smaller pieces, so a larger task like 'make a peanut butter sandwich', for example, becomes a sequences of steps that can easily be learned.

3. Get Them Active

Adults like to actively participate in learning experiences. Lecture, while important to learning, ought to be delivered in limited doses and reinforced by other learning methods. Demonstrations. Discussion. Question and Answer. Assessments. Case Studies.

4. Check Progress

How effective is the learning? More specifically, how effectively can the learners execute the learning objectives when the training is done? Evaluate progress, both during learning and at the end of it. The best place to evaluate progress? On the job. Learning that's applied to the job is the best indicator of success.

5. Give Feedback

Tell them when they've got it right. Correct them when they've got it wrong. And, remember what Mary Kay said...

"Get ordinary people to do extraordinary things. Give them something simple and when they do it, praise the hell out of them."

YOUR NOTES:

ABOUT CHRIS AND DEB

Christina M. Battell

Chris is the owner of Midwest Consulting Group, LLC providing coaching, training and consulting solutions to a variety of business and organizations. For the past 30+ years, Chris has developed and delivered solutions such as:

- Leadership Training and Coaching
- Conflict Resolution Training and Facilitation
- Management Development Training
- Team Building
- Performance Management Assessment
- Strategic Planning
- Event Facilitation

Email Chris: ChristinaMBattell@gmail.com

Deborah C. Miller, MA, MOQ/OE, PMP®

Deb is the owner of Miller Productivity, providing instructional design, project management, and procedural documentation services to financial services, healthcare, manufacturing and non-profit organizations. Deb is a Project Manager with

18+ years of experience leading projects and teams. She is a best-selling author and coach. Her services include:

- Instructional Design and Consulting for eLearning and blended courses.
- Procedure Analysis and Documentation
- Process Improvement
- Strategic Planning
- Change Management Assessments and Consulting

Email Deb at: DMiller@MillerProductivity.com

CONTACT CHRIS & DEB

For more information about our consulting services, contact us through our websites.

Chris: ChristinaBattell.com

Deb: MillerProductivity.com

If you missed *125 Wow! Change Cards: Tools For Leaders During Change*, you can order your copy from Amazon.

If you are interested in purchasing our books in bulk for use in your organization, send an email to dmiller@MillerProductivity.com and ask about our current rates.

Printed in Dunstable, United Kingdom

91220878R00218